AF320435

aus der Reihe:

Innovationen mit Mikrowellen und Licht

**Forschungsberichte aus dem Ferdinand-Braun-Institut,
Leibniz-Institut für Höchstfrequenztechnik**

Band 43

Martin Winterfeldt

Investigation of slow-axis beam quality degradation
in high-power broad area diode lasers

Herausgeber: Prof. Dr. Günther Tränkle, Prof. Dr.-Ing. Wolfgang Heinrich

Ferdinand-Braun-Institut
Leibniz-Institut
für Höchstfrequenztechnik (FBH)
Gustav-Kirchhoff-Straße 4
12489 Berlin

Tel.   +49.30.6392-2600
Fax   +49.30.6392-2602

E-Mail fbh@fbh-berlin.de
Web   www.fbh-berlin.de

Innovations with Microwaves and Light

**Research Reports from the Ferdinand-Braun-Institut,
Leibniz-Institut für Höchstfrequenztechnik**

Preface of the Editors

Research-based ideas, developments, and concepts are the basis of scientific progress and competitiveness, expanding human knowledge and being expressed technologically as inventions. The resulting innovative products and services eventually find their way into public life.

Accordingly, the *"Research Reports from the Ferdinand-Braun-Institut, Leibniz-Institut für Höchstfrequenztechnik"* series compile the institute's latest research and developments. We would like to make our results broadly accessible and to stimulate further discussions, not least to enable as many of our developments as possible to enhance everyday life.

High-power GaAs-based broad area (BA) diode lasers are widely used as pump sources for solid-state-based laser cutting machines in materials processing. The laser beam quality of these emitters has been identified as the key property ensuring high power densities for pumping and for direct diode applications. Hence, this work encompasses a detailed root cause analysis of the influences on BA laser lateral beam quality degradation, quantifying the most important factors. As a result, corresponding design concepts have been developed, enabling FBH to realize BA emitters that deliver power densities which exceed those of commercially available diode lasers.

We wish you an informative and inspiring reading

Prof. Dr. Günther Tränkle
Director

Prof. Dr.-Ing. Wolfgang Heinrich
Deputy Director

The Ferdinand-Braun-Institut

The Ferdinand-Braun-Institut researches electronic and optical components, modules and systems based on compound semiconductors. These devices are key enablers that address the needs of today's society in fields like communications, energy, health and mobility. Specifically, FBH develops light sources from the visible to the ultra-violet spectral range: high-power diode lasers with excellent beam quality, UV light sources and hybrid laser systems. Applications range from medical technology, high-precision metrology and sensors to optical communications in space. In the field of microwaves, FBH develops high-efficiency multi-functional power amplifiers and millimeter wave frontends targeting energy-efficient mobile communications as well as car safety systems. In addition, compact atmospheric microwave plasma sources that operate with economic low-voltage drivers are fabricated for use in a variety of applications, such as the treatment of skin diseases.

The FBH is a competence center for III-V compound semiconductors and has a strong international reputation. FBH competence covers the full range of capabilities, from design to fabrication to device characterization.

In close cooperation with industry, its research results lead to cutting-edge products. The institute also successfully turns innovative product ideas into spin-off companies. Thus, working in strategic partnerships with industry, FBH assures Germany's technological excellence in microwave and optoelectronic research.

INVESTIGATION OF SLOW-AXIS BEAM QUALITY DEGRADATION IN HIGH-POWER BROAD AREA DIODE LASERS

Vom Fachbereich Elektrotechnik der Technischen Universität Berlin
zur Erlangung der Würde eines
Doktors der Naturwissenschaften (Dr. rer. nat.)
genehmigte Dissertation

vorgelegt von

Martin Winterfeldt

aus Schwerin (Meckl.)

Erstgutachter: Prof. Dr. Günther Tränkle

Zweitgutachter: Prof. Dr. Eric Larkins

Drittgutachter: Prof. Dr. Michael Kneissl

Tag der mündlichen Aussprache: 14.09.2017

Ferdinand-Braun-Institut, Leibniz-Institut für Höchstfrequenztechnik

2017

Bibliografische Information der Deutschen Nationalbibliothek

Die Deutsche Nationalbibliothek verzeichnet diese Publikation in der Deutschen Nationalbibliografie; detaillierte bibliografische Daten sind im Internet über http://dnb.d-nb.de abrufbar.

1. Aufl. - Göttingen: Cuvillier, 2018

Zugl.: (TU) Berlin, Univ., Diss., 2017

© CUVILLIER VERLAG, Göttingen 2018

Nonnenstieg 8, 37075 Göttingen

Telefon: 0551-54724-0

Telefax: 0551-54724-21

www.cuvillier.de

Alle Rechte vorbehalten. Ohne ausdrückliche Genehmigung

des Verlages ist es nicht gestattet, das Buch oder Teile

daraus auf fotomechanischem Weg (Fotokopie, Mikrokopie)

zu vervielfältigen.

1. Auflage, 2018

Gedruckt auf umweltfreundlichem, säurefreiem Papier aus nachhaltiger Forstwirtschaft

ISBN 978-3-7369-9733-2

eISBN 978-3-7369-8733-3

Publications

Parts of this dissertation have already been published:

Journals

„Experimental investigation of factors limiting slow axis beam quality in 9xx nm high power broad area diode lasers",
M. Winterfeldt, P. Crump, H. Wenzel, G. Erbert and G. Tränkle
Journal of Applied Physics, **116**, 063103 (2014)

„High Beam Quality in Broad Area Lasers via Suppression of Lateral Carrier Accumulation",
M. Winterfeldt, P. Crump, S. Knigge, A. Maaßdorf, U. Zeimer and G. Erbert
IEEE Photonics Technology Letters, **27(17)**, 1809 (2015)

International conferences and workshops

„Experimental Investigation of Limits to Slow Axis Beam Quality in High Power Broad Area Diode Lasers",
M. Winterfeldt, P. Crump, H. Wenzel, G. Erbert and G. Tränkle
2014 International Semiconductor Laser Conference, Palma de Mallorca, pp. 193-194 (2014)

„The Influence of differential modal gain on the filamentary behavior of broad area diode lasers",
M. Winterfeldt, P. Crump, S. Knigge, A. Maßdorf, G. Erbert, G. Tränkle
2015 IEEE Photonics Conference (IPC), Reston, VA, pp.567-568 (2015)

„Limitations to brightness in high power laser diodes",
P. Crump, **M. Winterfeldt**, J. Decker, M. Ekterai, J. Fricke, A. Maßdorf, G. Erbert, G. Tränkle
2015 IEEE Photonics Conference (IPC), Reston, VA, pp.553-554 (2015)

„Assessing the influence of the vertical epitaxial layer design on the lateral beam quality of high-power broad area diode lasers",
M. Winterfeldt, J. Rieprich, S. Knigge, A. Maaßdorf, M. Hempel, R. Kernke, J.W. Tomm, G. Erbert, P. Crump
Proc. of SPIE, Vol. **9733**, High-Power Diode Laser Technology and Applications XIV, 97330O (2016)

„Novel approaches to increasing the brightness of broad area lasers",
P. Crump, **M. Winterfeldt**, J. Decker, M. Ekterai, J. Fricke, S. Knigge, A. Maßdorf, G. Erbert
Proc. of SPIE, Vol. **9767**, Novel In-Plane Semiconductor Lasers XV, 97671L (2016)

„Assessing the impact of thermal barriers on the thermal lens shape in high power broad area diode lasers",
J. Rieprich, **M. Winterfeldt**, J.W. Tomm, P. Crump
2016 International Semiconductor Laser Conference (ISLC), Kobe, pp. 1-2 (2016)

„Assessment of factors regulating the thermal lens profile and lateral brightness in high power diode lasers",
J. Rieprich, **M. Winterfeldt**, J.W. Tomm, R. Kernke, P. Crump
Proc. of SPIE, Vol. **10085**, Components and Packaging for Laser Systems III, 1008502 (2017)

Abstract

Modern High Power Broad Area Diode Lasers (BAL) are an important building block of the materials processing industry, exhibiting high optical output power P_{opt} with high power conversion efficiency η_c. However, their use in direct diode systems, e.g. for sheet metal cutting, is limited due to poor lateral (in-plane) beam quality, parameterized via the beam parameter product $BPP_{lat} = 0.25 \times \Theta_{95\%} \times w_{95\%}$, with $\Theta_{95\%}$ and $w_{95\%}$ denoting the full divergence angle and the beam waist width at 95% power content, respectively. High BPP_{lat} limits the power density of focused BAL radiation, an important industry measure, expressed via the linear radiance $B_{lin} = P_{opt}/BPP_{lat}$.

The main objective of this doctoral thesis is an analysis of the factors that influence BPP_{lat}. Therefore, a series of diagnostic experiments is conducted, each aiming on a specific potential influence in order to assess its importance. The list of considered effects encompasses the thermal lens shape, the epitaxial laser design, the lateral carrier profile, process induced index guiding via dry etched trenches, filamentation and mechanical strain. The analysis is supported by a simple linear model of the BPP_{lat} growth with active zone temperature increase ΔT_{AZ} that is introduced as an analytic tool: $BPP_{lat}(\Delta T_{AZ}) = BPP_0 + S_{th} \cdot \Delta T_{AZ}$. Here, the ground level BPP_0 and the thermal slope S_{th} are important indicators for successful BPP_{lat} reduction. The analysis revealed that the epitaxial layer design and the chip geometry have a considerable impact on the thermal lens bowing, which is directly correlated to the thermal slope S_{th}. In addition, the suppression of lateral carrier accumulation with deep proton implantation at the BAL emitter edges led to a 33% decrease in thermal slope, revealing that one third of S_{th} is regulated by the current driven gain supply to higher order modes, while the remaining two thirds is regulated by the thermal lens shape. Proton implanted devices show good linear radiance of $B_{lin} = 3.5\,\text{W/mm}\,\text{mrad}$, but suffer from an implantation induced efficiency loss of 7%-points. The application of index guiding trenches with an effective index step of $\Delta n_{eff} = 1.5 \times 10^{-3}$ yielded stabilized near field dimensions as function of current, but also a 1.5 mm mrad increase in BPP_0. Filamentation and mechanical strain were found to marginally affect BPP_{lat} in modern strained quantum well BALs.

The second part of this thesis is devoted to the assessment of techniques that aim to reduce BPP_{lat} via a reduction of the number of active lateral modes. First, measurements on BALs with varying stripe width w showed increased efficiency and higher output power for broad injection stripes. However, the growth in output power is overcompensated by a growth in BPP_{lat}, yielding for highest radiance in narrow stripe ($w = 30\,\mu m$) BALs with $B_{lin} = 4\,W/mm\,mrad$ at $P_{opt} = 4\,W$. A simulation of thermal waveguides showed that the BPP_{lat} deterioration rate can be emulated with simple uniform mode overlap and amounts to $\Delta BPP_{lat}/\Delta w = (0.044 \pm 0.002)\,mm\,mrad/\mu m$ due to an increase in the number of guided lateral modes with increasing w (fixed thermal waveguide with $\Delta T_{AZ} = 15\,K$ assumed). This result is in good agreement with the measured BPP_{lat} increase for stripe widths up to $w = 70\,\mu m$. Second, a lateral mode filter approach based on the resonant coupling of two congruent vertical waveguides was tested. Here, the out-coupling waveguide (germanium filled, dry etched trenches beside injection stripe) exhibits strong optical absorption that introduces a near-field-width-selective loss mechanism for the modes in the central (lasing) waveguide. The conversion efficiency is heavily compromised by the mode filter and is limited to $\eta_c \leq 45\%$, leaving room for design optimizations. However, the mode filter showed promising results in terms of linear radiance, yielding $B_{lin} = 4.4\,W/mm\,mrad$, which exceeds the radiance of commercially available broad area lasers.

Kurzfassung

Moderne Breitstreifen-Hochleistungsdiodenlaser (BALs) sind ein wichtiger Grundbaustein der material-verarbeitenden Industrie, denn sie erreichen hohe optische Ausgangsleistungen P_{opt} bei gleichzeitig hoher Konversionseffizienz η_c. Dennoch sind sie für die direkte Anwendung, zum Beispiel beim Laser-Schneiden, ungeeignet, da sie eine geringe laterale (senkrecht zur Wachstumsrichtung und zur Achse der Lichtausbreitung) Strahlqualität aufweisen. Letztere wird durch das Strahlparameterprodukt quantifiziert: $BPP_{lat} = 0.25 \times \Theta_{95\%} \times w_{95\%}$, wobei $\Theta_{95\%}$ den vollen Divergenzwinkel und $w_{95\%}$ die volle Strahltaille bei 95% Leistungsinhalt beschreiben. Ein hohes Strahlparameterprodukt limitiert die Leistungsdichte, die ein BAL erreichen kann. Diese Leistungsdichte ist eine wichtige Kennzahl für industrielle Lasersysteme und wird durch die lineare Brillanz $B_{lin} = P_{opt}/BPP_{lat}$ ausgedrückt.

Das Hauptziel dieser Doktorarbeit ist eine Analyse der Einflussfaktoren auf die laterale Strahlqualität BPP_{lat}. Hierzu wird eine Reihe von Diagnose-Experimenten durchgeführt, die jeweils darauf zugeschnitten sind, den Einfluss eines bestimmten Effekts zu untersuchen. Im Rahmen dieser Arbeit werden auf diese Weise die folgenden Faktoren untersucht: die Form der thermischen Linse, die Epitaxiestruktur, das laterale Ladungsträgerprofil, trocken geätzte Indexgräben, Filamentierung und mechanische Verspannungen. Zur Analyse wird dabei ein einfaches empirisches Modell eingeführt das die Veränderung von BPP_{lat} mit der Erwärmung der aktiven Zone ΔT_{AZ} als linearen Anstieg beschreibt: $BPP_{lat}(\Delta T_{AZ}) = BPP_0 + S_{th} \cdot \Delta T_{AZ}$. Hierbei stellen der Ordinatenabschnitt BPP_0 (BPP 'Grundlevel') und der thermische Anstieg S_{th} zwei wichtige Indikatoren für die Veränderung des Strahlparameterprodukts BPP_{lat} dar. Die Analyse ergab, dass die Epitaxiestruktur im Speziellen und die Laserchip-Geometrie im Allgemeinen einen starken Einfluss auf die Form der thermischen Linse haben, wobei Letztere direkt mit S_{th} korreliert. Zusätzlich zeigte sich, dass eine Unterdrückung der lateralen Ladungsträger-Ansammlung mit Hilfe von tiefer Protonenimplantation den thermischen Anstieg um 33% senkt, so dass ein Drittel des Betrags von S_{th} durch den vom Pumpstrom induzierten Gewinn für Moden höherer Ordnung bestimmt wird und die verbleibenden zwei Drittel durch die Form der thermischen Linse bestimmt werden. Die implantierten BALs zeigten eine lineare Brillanz von $B_{lin} = 3.5\,\text{W/mm\,mrad}$, erleiden durch die Implantation aber einen Effizienzverlust von 7 Prozentpunkten. Die Anwendung von Indexgräben mit einem effektiven Indexsprung von $\Delta n_{eff} = 1.5 \times 10^{-3}$ führte zu einer Stabilisierung der Nahfeldbreite bei steigenden Pumpströmen, aber gleichzeitig wurde das BPP Grundlevel BPP_0 um 1.5 mm mrad angehoben. Filamentierung und mechanische Verspannungen zeigten in dieser Untersuchung nur einen minimalen Einfluss auf das Strahlparameterprodukt.

Im zweiten Teil dieser Arbeit wurden Techniken untersucht, die BPP_{lat} über eine Reduzierung der Anzahl lateraler Moden verbessern. Hierbei wurden zunächst BALs mit unterschiedlicher Breite des Injektionsstreifens w untersucht, wobei Laser mit großen Streifenbreiten höhere Effizienzen und höhere optische Leistungen zeigten. Da das Strahlparameterprodukt jedoch schneller mit w wächst als die Ausgangsleistung, wird die höchste Brillanz in BALs mit schmalen Streifen ($w = 30\,\mu\text{m}$) erzielt: $B_{lin} = 4\,\text{W/mm\,mrad}$ bei $P_{opt} = 4\,\text{W}$. Eine Simulation der thermischen Wellenleiter (bei festem $\Delta T_{AZ} = 15\,\text{K}$) ergab, dass die Degradationsrate von BPP_{lat} mit einer einfachen, gleich-gewichteten Überlagerung der geführten Moden nachgebildet werden kann.

Die ermittelte Rate $\Delta BPP_{lat}/\Delta w = (0.044 \pm 0.002)\,\text{mm\,mrad/\mu m}$ wird durch den Zuwachs in der Anzahl geführter Moden mit zunehmender Streifenbreite bestimmt und stimmt sehr gut mit den gemessenen BPP_{lat}-Werten bis zu $w = 70\,\mu\text{m}$ überein. Die zweite untersuchte Technik ist ein lateraler Modenfilter, der auf der resonanten Kopplung zweier kongruenter vertikaler Wellenleiter basiert. Hierbei gibt es einen Auskoppel-Wellenleiter (realisiert durch trocken geätzte, mit einer Germaniumschicht versehene, Indexgräben neben dem Injektionsstreifen) der eine starke optische Absorption aufweist und einen Verlust für die lateralen Moden des Zentralbereichs (Laserbereich) einführt, dessen Stärke von der Nahfeldausdehnung der Moden abhängt. Der Einsatz des Modenfilters führt zu starken Einbußen in der Konversionseffizienz, die auf $\eta_c \leq 45\%$ beschränkt ist, so dass die Technologie als Ganzes noch Verbesserungspotential aufweist. Dennoch zeigten die BALs mit Modenfilter vielversprechende Brillanz-Ergebnisse, mit einem Bestwert von $B_{lin} = 4.4\,\text{W/mm\,mrad}$. Dieses Ergebnis übersteigt die Brillanzwerte, die von kommerziell verfügbaren Breitstreifenlasern erreicht werden.

Acknowledgments

First of all, I would like to express my sincere gratitude to Prof. Dr. Günther Tränkle for giving me the opportunity to carry out my PhD studies at the Ferdinand-Braun-Institut, Leibniz Institut für Höchstfrequenztechnik (FBH). In every consultation he gave good advice and constructive critique and helped me in finding solutions to the problems I encountered. My thanks also go to Prof. Dr. Eric Larkins and Prof. Dr. Michael Kneissl who dedicated their time for an external review of my PhD thesis.

I am indebted to Dr. Paul Crump, head of the High-Power Diode Laser Lab at the FBH, for his guidance, his support and his unbreakable enthusiasm. Paul was a mentor for me and taught me that no matter how confusing a data set looks, that 'somewhere in there is truth and beauty'. Well - truth, beauty *and* unforeseen gremlins, of course.

How good you feel at work strongly depends on the people who are your colleagues. My memory of my time at the FBH will always be connected to my PhD-fellows: Marcel Braune, Norman Ruhnke, Jonathan Decker, Carlo Frevert, Thorben Kaul, Matthias Karow and Juliane Rieprich. I cannot thank you enough! Without you guys, my work-life would have been much more painful and tedious.

I warmly thank my other FBH-colleagues for supporting my work. Especially Dr. Gunnar Blume and Ralf Staske for their technical advice. Furthermore, I am sincerely grateful to Dr. Jörg Fricke, Dr. Steffen Knigge, Dr. Peter Ressel, Dr. Andre Maaßdorf and all other colleagues, who have contributed to the epitaxy, processing and mounting of the diode lasers. Special thanks go to Dr. Ute Zeimer and Helen Lawrenz for their support in the SEM-analysis of deeply implanted emitters.

I further thank my parents and my brothers Thomas and Stefan for their love and encouragement. In addition, I am greatly indebted to my friends, whose support I always received when I needed it most: Christian, Lene, Yves, Anni, Felix, Lisa, Robert, Steffi, Ole, Peer, Silvia and Kerstin. I am glad to know you.

The financial support by Trumpf Photonics (TUSP) is gratefully acknowledged.

Contents

Chapter 1

Introduction

Throughout the last five decades, the semiconductor diode laser has developed into a technology of great importance to our modern society. In terms of performance, robustness and diversity, this field has experienced consistent progress, driven by academic and industrial research. Nowadays diode lasers find applications in fiber based telecommunications ('fast' internet connections), materials processing, surgery, gas- and ranging sensors and household electronics (e.g. blu ray player, laser printer). Moreover, their high reliability enables the use of diode lasers in places that are considered rather remote, e.g. in optical amplifiers in deep-sea cables, in communication units of satellites or in sensors of Mars exploration rovers.

The object of study in this doctoral thesis is the gallium-arsenide based, high power *broad area laser* (BAL), an opto-electronic device that had its advent in the early 1960s and has since developed into a key technology for laser based materials processing. Driven by an electric current, it produces laser emission in a wide range of available wavelengths $630\,\mathrm{nm} \leq \lambda \leq 1120\,\mathrm{nm}$, exhibiting long operational lifetime and high output powers at conversion efficiencies $> 60\%$. However, in terms of *beam quality*, the BAL performance is poor compared to competing laser systems such as the disk- or fiber lasers, limiting its usage in direct diode systems. Hence, motivated by a growing industry demand for efficient, high brightness laser sources, the focus in this work is placed on understanding the physical limitations to the BAL's beam quality.

This chapter starts with a motivation that includes an overview of the current laser market, followed by a comparison of state-of-the-art materials processing laser systems and a presentation of recent achievements in BAL performance. Afterwards, the methodology of this thesis will be presented and finally the contents of this thesis will be briefly summarized.

1.1 Motivation - High Power Diode Lasers in materials processing

Figure 1.1 illustrates the commercial relevance of diode lasers. In 2016, total diode laser revenues reached US $ 4.68 billion, which is 45% of the entire laser market. Moreover, figure 1.1(a) shows that the laser market has been expanding at an average growth rate of 5% p.a. within the last 3 years and the market share for diode lasers is relatively stable at $\approx$ 45.5%. The origin of this strong industry demand for diode lasers lies in a variety of beneficial properties, most notably compactness, easy integration into electronic circuits, wavelength tunability and reliability.

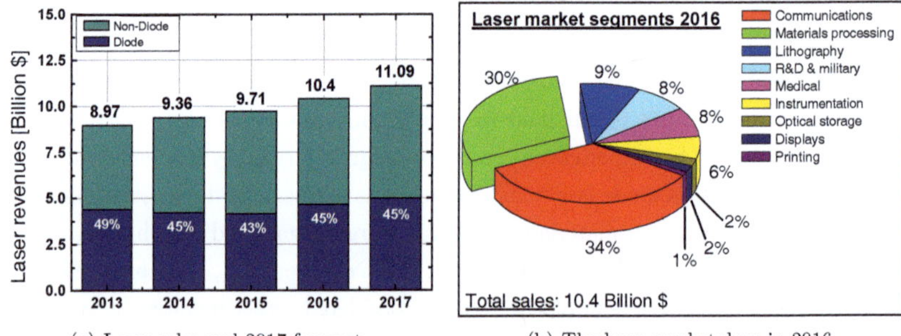

(a) Laser sales and 2017 forecast. (b) The laser marketplace in 2016.

Figure 1.1: Current status of the international laser market, data taken from [1]. Left: Total laser revenues in billion US $ of the past 4 years and forecast for 2017 shows a growing market (average growth rate of 5% p.a. in the last 3 years). The share of diode lasers on the whole market is stable at $\approx$ 45.5% (4-year average). Right: Illustration of the laser market segments shows 34% of the laser revenues in 2016 belongs to the communications sector and another 30% to materials processing. All other laser industries share the remaining 36%.

The pie chart in figure 1.1(b) shows that among the laser market segments in 2016, materials processing is the second largest industry with 30% market share, closely following the communications sector (34%).

Figure 1.2 also shows that sheet metal cutting is the most widespread application in laser based materials processing, with a lion's share of 36% of the laser revenues in that field. Now, in materials processing in general, but especially in sheet metal cutting, the high power broad area laser has an important function as building block for the laser systems that are used here.

BALs show output powers in continuous wave (cw) mode of $P_{opt} > 20\,\text{W}$ and peak power conversion efficiencies of $\eta_c \approx 70\%$. In addition, the assembly in a wafer fab allows low-cost mass production, enabling a significant reduction of the purchase price in terms of US $

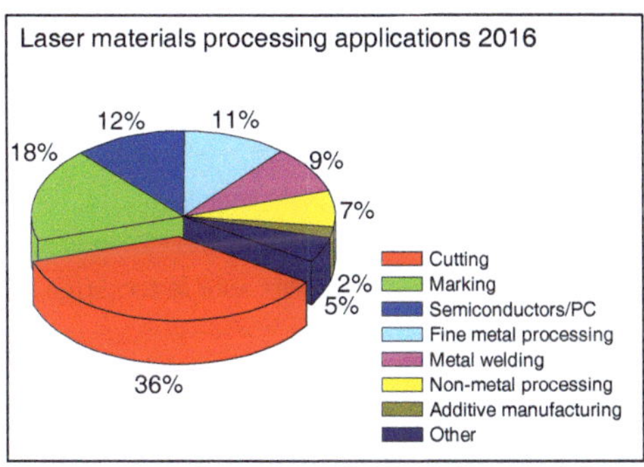

Figure 1.2: The materials processing field divided into relevant applications [1].

per watt of useful optical power (industry measure '\$/W'). This makes the BAL the ideal supplier of pump light for solid-state laser systems, e.g. Yb/Nd:YAG-, disk- and fiber lasers, which are used for deep penetration welding and sheet metal cutting. The latter two applications demand very high power densities of laser radiation, parameterized via the *radiance* $B = P_{opt}/\pi^2 \cdot BPP^2$. This is the formula for rotationally symmetric beams, the definition of B will be introduced in detail in section 2.1.4. For sufficient radiance B, high output powers in the kW-range and more importantly a low *beam parameter product* ('BPP' = beam waist radius × half divergence angle, cf. equation 2.7 in section 2.1.4) are needed.

Disk- and fiber lasers reach sufficient power densities for sheet metal cutting and, due to the efficient BAL pumping, their wall-plug efficiency ($\eta_{wp} \approx 30\%$) easily exceeds that of CO_2-lasers ($\eta_{wp} \approx 12\%$). However, since the energy delivered to the workpiece is converted in two steps, from electrical energy to pump light and then to 'high-beam-quality' (i.e. low BPP) light, the wall-plug efficiency of these systems is still low compared to *direct diode systems*. The latter reach wall-plug efficiencies $\eta_{wp} > 40\%$ [2] and power levels in the kW-range through beam combination of multiple BALs in fiber-coupled modules, using polarization- and wavelength multiplexing techniques. However, their radiance is too low for use in sheet metal cutting, due to their high beam parameter product.

This problem is illustrated in figure 1.3, where the required laser beam properties of common applications in materials processing are marked in a BPP vs. output power map. Here, the radiance increases from the top left corner to the bottom right corner.

The performance of diode laser systems is shown as an orange line and all accessible applications are depicted in light-grey. Today, direct diode systems with 8 kW optical output power and BPP of 7 mm mrad are commercially available (cf. table 1.1). However, as indicated by the slope of the orange line, high output powers can only be reached at the cost of high BPP. Hence, fiber lasers (1 kW at 0.37 mm mrad, blue rectangle) and disk lasers (8 kW at 4 mm mrad, yellow dots) still prevail in the lucrative markets for sheet metal cutting and deep welding.

The origin of the radiance deficit in direct diode systems is the lateral (slow-axis) beam quality of broad area lasers. In the lateral direction, the emission is multi-moded and far away from diffraction limited, such that the focus spot is too large to allow coupling into low-NA fibers.

The increased slow-axis BPP in broad area lasers reveals the motivation of this doctoral thesis. A more detailed understanding of the physical limitations and contributing effects to this problem is necessary. Hence, the main objective of this thesis is a root cause analysis that seeks to identify the most influential effects on the BALs beam parameter product. Ideally, the broadened knowledge from this investigation can be used to suggest sophisticated lateral laser designs that will improve the BALs beam quality without compromising output power, efficiency and reliability. This will enable a higher radiance for energy-efficient direct diode systems, which are in strong demand in the materials processing sector.

In order to gain a more detailed picture of the current status of industrial laser systems and BAL performance, a state-of-the-art overview will be presented in the following section.

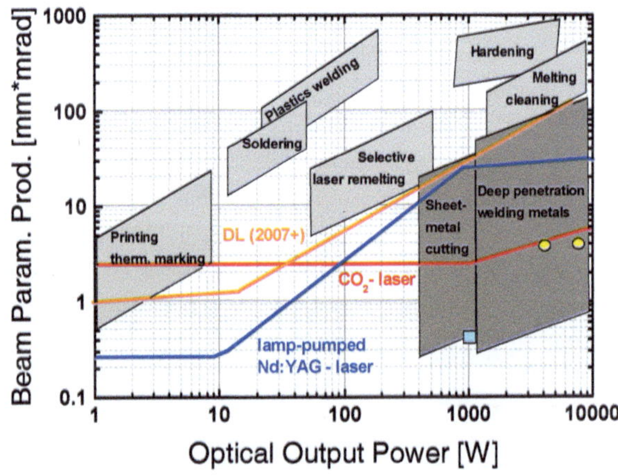

Figure 1.3: Materials processing laser 'roadmap', adapted from [3]. Beam parameter product (BPP) plotted versus optical output power in double-logarithmic scale. The laser source radiance B increases at higher powers and reduced BPP. Relevant laser applications and performance of different laser species are noted. Today the highest radiance is achieved by disk lasers (Trumpf 'TruDisk' series, yellow dots) and fiber lasers (IPG 'YLR-1000WC', blue rectangle).

Review of state-of-the-art diode laser technology

A list of commercially available laser systems for materials processing is shown in table 1.1, ordered in descending values of linear radiance $B_{lin} = P_{opt}/BPP$. As already illustrated in figure 1.3, direct diode systems - though very efficient - exhibit inferior beam quality, i.e. they reach high output powers only at cost of a high beam parameter product. Hence, most of these systems occur at the end of the list. At the top, however, fiber-, CO_2- and disk lasers show a linear radiance $B_{lin} \geq 2\,\text{kW}\,\text{mm}^{-1}\,\text{mrad}^{-1}$.

B_{lin} [*]	P_{opt} [kW]	BPP [mm mrad]	λ [µm]	η_c	type	vendor/product
2.7	1	0.37	1.07	0.31**	Yb fiber	IPG YLR-1000WC
2.3	8	3.5	10.6	-	CO_2	Rofin DC series
2	16	8	1.03	>0.3	Yb:YAG disk	Trumpf TruDisk 16002
1.6	10	6.2	10.6	0.15	CO_2	Trumpf TruFlow 10000
1.1	8	7	0.97	-	direct diode	Teradiode
1	8	8	1.07	0.4	Yb fiber	IPG YLS-CUT
0.18	11	60	0.9-1.07	0.36**	direct diode	Laserline LDF series
0.15	1.5	10	0.976	0.33**	direct diode	IPG DLR-976-1500
0.13	4	30	0.92-1.04	0.42	direct diode	Trumpf TruDiode4006

Table 1.1: Commercially available industry laser systems for materials processing, listed in descending linear radiance $B_{lin} = P_{opt}/BPP$, measured in (*) $\text{kW}\,\text{mm}^{-1}\,\text{mrad}^{-1}$. Data taken from online product catalogs (03/2017).(**) efficiency calculated based on noted power consumption. (-) Data not available.

Since the main objective of this thesis is the broad area laser, a look at the current status of BAL performance in the 9xx nm wavelength range is worthwhile. Here, two lists are presented: The first (cf. table 1.2) shows an overview of commercially available broad area lasers and laser bars, respectively. The second list (cf. table 1.3) shows the latest research results in the high brightness diode laser field. It should be noted that both lists do not claim completeness. In addition, the supply of information is quite different among research groups and diode laser vendors. Hence, not all parameters are available for all of the listed devices.

In comparison to the B_{lin}-values for direct diode systems listed in table 1.1, the radiance of single emitters is reduced by approximately two orders of magnitude. The radiance up-scaling in direct diode systems is usually done by dense and coarse wavelength beam combining (WBC) as well as polarization multiplexing. The latter ideally increases the radiance by a factor of x2, while WBC scales the radiance with the number of wavelength channels N_λ and an efficiency factor f_{WBC} [4]. However, since the BAL is the basic element

B_{lin}	P_{opt}	BPP_{lat}	w	Θ	η_c	comment	vendor
[*]	[W]	[mm mrad]	[µm]	[°]			
3.5	11	3.14	90	8	>0.55	FWHM, 975 nm	II-VI
2.6	8**	3.05	100	7	0.65	95% p.c., 980 nm	Osram OS
1.4	4.2**	3.05	100	7	0.62	95% p.c., 938 nm	JDL
1.3	4.2**	3.14	90	8	-	FWHM, 975 nm	coherent

Table 1.2: Commercially available 9xx nm broad area emitters and laser bars, listed in descending linear radiance $B_{lin} = P_{opt}/BPP_{lat}$, measured in (*) $\mathrm{W\,mm^{-1}\,mrad^{-1}}$. Data taken from online product catalogs (03/2017).(**) Power per emitter in laser bar. (-) Data not available.

of a direct diode system, the overall radiance is increased with improved BAL radiance. The brightest commercially available diode laser emitters reach 2-3.5 $\mathrm{W\,mm^{-1}\,mrad^{-1}}$. In terms of conversion efficiency, these emitters reach values $\eta_c > 60\%$, indicating that a considerable amount of useful output power is lost in direct diode beam combining systems where $\eta_c \approx 40\%$.

B_{lin}	P_{opt}	BPP_{lat}	w	Θ	η_c	comment	reference
[*]	[W]	[mm mrad]	[µm]	[°]			
4.8	-	-	-	-	-	press release	OSRAM [5]
4.4	11	2.5	90	7	0.45	960 nm AWL-BAL	this work
4.3	13	3	-	7	0.63	9xx nm	nLight [6]
3.5	7	2	90	7	0.53	95% p.c., 969 nm	this work, [7]
3.2	8	2.5	-	-	-	9xx nm	Eckstein [8]
3.15	10.4**	3.3	100	7	0.69	95% p.c., 976 nm	OSRAM [9]
2.8	11**	4	90	10.2	0.69	95% p.c., 915 nm	JDL [10]

Table 1.3: Recent development in 9xx nm BALs/laser bars, listed in descending linear radiance $B_{lin} = P_{opt}/BPP_{lat}$, measured in (*) $\mathrm{W\,mm^{-1}\,mrad^{-1}}$. (**) Power per emitter in laser bar. (-) Data not available.

The results of recent research activities are presented in table 1.3, showing a considerable increase in BAL radiance compared to commercially available standards. The last two entries are results from 5-emitter minibars, showing very high conversion efficiencies of $\eta_c = 69\%$. Two results were obtained in the course of this doctoral thesis. First, deeply implanted $w = 90\,\mu m$ BALs with suppressed lateral carrier accumulation show a linear radiance of $B_{lin} = 3.5\,\mathrm{W\,mm^{-1}\,mrad^{-1}}$ at $P_{opt} = 7\,W$ (cf. section 3.4). Secondly, BALs with a lateral mode filter ('*anti-waveguide*' layer, AWL) reach $B_{lin} = 4.4\,\mathrm{W\,mm^{-1}\,mrad^{-1}}$ at $P_{opt} = 11\,W$ (cf. section 4.2). However, both improvements compromise the power conversion efficiency, leaving space for optimization.

1.2 Method - Root cause analysis

The usual way to expand the knowledge on an object of study is a stepwise progress that contains the creation of a (simplified) model of the interaction of physical parameters, the implementation of this model in simulations and the subsequent comparison of predicted outcomes to measurements. However, in broad area lasers a complex interaction of the many internal parameters (refractive index, temperature, carrier density, gain, electric field, material strain etc.) exacerbates thorough device simulation, such that different models can lead to consistency with the measured data and no clear answer on the limiting effects can be given. Hence, the focus is shifted to experimental methods that aim for increased device performance (PDCA-cycling) and on understanding of the underlying effects (root-cause-analysis).

The main progress in diode laser technology is based on a simple iteration-cycle:

- **Plan**: Develop laser design on basis of present knowledge

- **Do**: Device fabrication in wafer fab (lithography, etching etc.)

- **Check**: Characterization of multiple emitters in parameter-matrices

- **Act**: Feedback to laser design and fabrication chain.

This PDCA-cycle is industry standard and appropriate to improve existing technologies by repeating the iterations until a performance goal is reached. However, in the well established diode laser technology, where the learning curve saturates, the PDCA-cycle is often insufficient to reach the desired performance. Due to the complex and comprehensive fabrication process many potential influences arise. Hence, it is more and more difficult to achieve reproducibility and to correlate incremental changes in laser performance to specific design measures.

Furthermore, a PDCA cycle does not necessarily yield results that provide more *understanding* of the underlying effects. But for a further technology improvement, this understanding is crucial! It makes it possible to estimate fundamental limits of important laser properties and enables the development of more accurate simulation-tools. In this way, new strategies for improved laser designs can be deduced that - ideally - exceed the progress made with PDCA-cycling.

Hence, in this thesis an alternative approach is adopted: the *root cause analysis*. Therein, based on the available knowledge, a list of potentially important effects is generated. Then an experiment is designed, such that each effect is isolated (if possible) in order to assess its contribution to the figure of interest. In this thesis, the BALs are designed to differ in only one property that is suspected to influence the lateral beam quality. In this way, 'diagnostic' series of laser devices are produced that give insights into the relative contributions of different limiting effects.

It should be noted that PDCA-cycling and root cause analysis are not competing strategies. On the contrary: they complement each other, so that progress in laser device performance can be reached faster. The root cause analysis answers the question '*What should we focus on?*', addressing device physics to identify limiting effects, while PDCA-cycling is inevitable to get the 'fine-tuning' of the relevant parameters, i.e. reaching the best performance based on the known limits.

1.3 Structure of this work

This doctoral thesis is divided into four main chapters that contain the following information:

In chapter 2, the theoretical background for the understanding of the results is provided. A brief introduction to the topic is followed by a revision of the basic characteristics of modern high power broad area lasers. The figures of *beam quality* and *radiance* are introduced and the problem of slow-axis beam quality degradation is outlined. Furthermore the most important techniques used are explained briefly, including the simulation of lateral modes in a thermal waveguide and the processing technique of ion implantation. Finally the measurement equipment and its accuracy will be presented.

The above mentioned root cause analysis is the core of this thesis and is presented in chapter 3, starting with a listing of all effects that are anticipated to influence the BAL's lateral beam quality. Next, a simple empirical model of BPP growth with the (local) active zone temperature is introduced, enabling the comparison of BALs with different thermal resistance and conversion efficiency. The following three main sections contain the results of diagnostic studies on BALs that focus on: the vertical design and its correlation to the thermal lens, process- and packaging induced effects on the BPP and the degree of polarization and the influence of the lateral carrier/gain profile. At the end of chapter 3, the most important findings are summarized and strategies for improved lateral beam quality are proposed.

Chapter 4 contains further results of experiments that are explicitly aimed at reducing the number of lateral modes. While in chapter 3 the BAL stripe width is fixed at $w = 90\,\mu m$, the first section in chapter 4 assesses changes in the behavior of BALs as function of stripe width. Here, the focus is on output power, efficiency, lateral beam quality and radiance as a function of injection stripe width. In the second section, the results of a lateral mode filter approach are presented. The so-called *'anti-waveguide'* generates high losses for modes with large lateral extent via a neighboring waveguide structure that is congruent to the central vertical laser waveguide but possesses a thin layer with strong optical absorption.

Finally, chapter 5 contains a summary of the most important observations in this thesis and a discussion of their generic validity. Furthermore, an outlook on possible further efforts for radiance enhancement in broad area lasers will be given.

Chapter 2

High Power Diode Lasers Fundamentals

In this chapter the main theoretical concepts are introduced. The groundwork for the understanding of the results is laid, assuming a basic knowledge on semiconductors and lasers. This chapter is structured as follows: It starts with a general introduction to the topic and a brief historical review. Then, in the first section, basic principles are repeated for convenience, including a presentation of broad area lasers as the main optoelectronic device in this study and a definition of beam quality and radiance. The second section encompasses the explanations of various effects and techniques that were used in this study. First, for optical simulation, an understanding of how the refractive index is influenced by various physical processes is necessary. Then, the simulation of the electric field requires the solution of the Helmholtz equation, which is presented here in the simple one dimensional case, restricting the analysis to the lateral coordinate. Secondly, the impact of proton implantation on the gallium arsenide crystal structure is outlined, as this technique will be used to shape the diode laser lateral carrier profile. Finally, the measurement equipment will be presented, including a discussion of its limitations.

2.1 Introduction to semiconductor diode lasers

Diode laser technology is a wide field and its study requires knowledge from different branches of physics such as solid state physics, quantum mechanics and (laser-)optics. In addition a basic understanding of semiconductor device properties and fabrication technologies is required. However, this work cannot encompass a complete synopsis of the background knowledge. Instead the reader is directed to the key references:

2.1.1 Introductory literature about diode lasers

Semiconductors, elementary or compound, are in the solid aggregate state at room temperature. They have a crystalline structure which determines their mechanical, electrical and optical properties. The formation of energy-bands for example is a direct consequence of the periodicity of atomic potentials in the semiconductor crystal. These topics are covered for example in the books by N. W. Ashcroft and N. D. Mermin [11] and C. Kittel [12].

Diode lasers are (coherent) light sources and the manipulation of light beams is described in the field of optics. Among the books on optics the *'Bergmann-Schaefer: Optik'* [13] is probably the most comprehensive compendium. An overview on the different laser technologies (solid state, gas, semiconductor) and their working principles is presented in the book of A.E. Siegman [14].

On semiconductor devices in general (including transistors, CCD-chips etc.) the standard reference is S. M. Sze and Kwok K. NG [15]. It leads the reader from basic building blocks to the description of important electronic and opto-electronic semiconductor devices. However, a great variety of books is available - some of them with a focus on fabrication technology [16] or semiconductor physics [17].

The reference [15] already contains a chapter on diode lasers, but this topic is comprehensive enough to fill its own literature list. The book written by G.P. Agrawal and N.K. Dutta [18] is a compendium that summarizes the state-of-the-art up to 1986 and gives insights to the physical effects that determine diode laser performance. However, for the description of the physics in diode lasers the books written by L.A. Coldren and S.W. Corzine [19] and S.L. Chuang [20] should be preferred, both having chapters that describe the physics on a phenomenological level as well as chapters that explain crucial methods in more detail. For the opto-electronic device engineer the book from P. Epperlein [21] is recommended, since it presents concrete laser devices and discusses their performance as function of their design parameters.

Finally two references cover high power broad area lasers. In the book editored by R. Diehl [22] the focus is laid on high power diode lasers, their fabrication and properties. The book written by F. Bachmann [3] covers high power single emitters and bars, their cooling and packaging as well as the construction of beam combining systems.

A short history of LASERs

Judged by its impact on science and technology, the invention of the LASER (**L**ight **A**mplification by **S**timulated **E**mission of **R**adiation) belongs to the most influential achievements in the 20th century. Its advent dates back to 1916, the year in which Albert Einstein introduced the process of *stimulated emission* in his approach to derive Planck's quantum hypothesis $E = h\nu$ by using rate equations for photon absorption and emission [23]. This process allows photon amplification provided that its probability exceeds that of photon absorption (*population inversion*). This is the fundamental physical effect that underlies every LASER system, but it took 42 years until Schawlow and Townes proposed a technique to produce amplified light emission in the infrared and optical frequency range in 1958 [24]. Shortly after that, in 1960, the first LASER in the visible wavelength range was built by T. Maiman, using a flash-lamp pumped ruby crystal [25].

The success story of diode lasers starts only two years after this major breakthrough. In 1962, Hall *et al.* observed coherent emission from a forward-biased GaAs p-n junction [26]. The electron-hole recombination in the depletion region provided the optical gain and the cleaved facets perpendicular to the junction plane provided the optical feedback. However, the early devices were impractical due to their enormous threshold current densities ($j_{thr} >$ $50\,\mathrm{kA\,cm^{-2}}$), so that continuous wave operation at room temperature was impossible. With the invention of the double-heterostructure [27], a significant improvement in carrier- and optical confinement was achieved, which led to the first injection lasers that operated in cw-mode at room temperature in 1969 [28]. In the following decades, continuous improvement of the technology was achieved by increased material quality and more sophisticated vertical designs. Furthermore a broader wavelength range was covered by the implementation of laser devices using other direct bandgap semiconductors, such as InAs, InP and GaAsP. In the early 1980s, the advent of the quantum-well diode laser [29] brought further significant improvements in threshold current density, differential efficiency and temperature stability.

The diode lasers in this work stem from the Ferdinand-Braun-Institut's wafer fab, which profits from more than two decades of practical experience in metal-organic vapor phase epitaxy (MOVPE) and the fabrication of quantum-well high power broad area lasers.

2.1.2 Modern High Power Broad Area Lasers

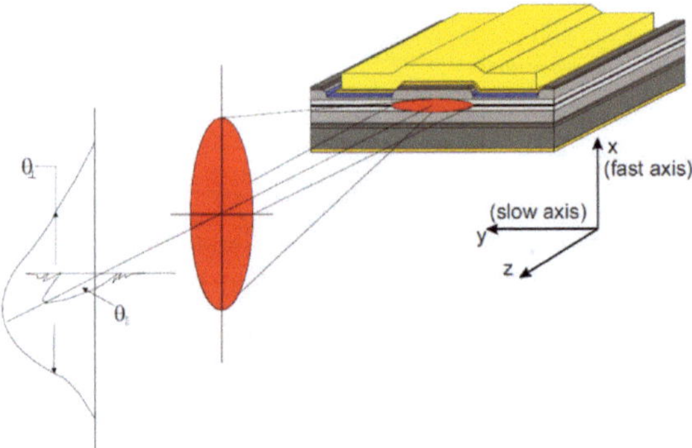

Figure 2.1: Schematic figure of a broad area laser, including light emission cone and common coordinate system, adapted from [3].

In figure 2.1, a broad area laser (BAL) is depicted in a schematic drawing. Based on continuous improvements in terms of carrier- and optical confinement, the modern BAL exhibits a sophisticated epitaxial layer design in the $Al_xGa_{1-x}As$ material system that is beyond a simple p-n-junction. In this study, the so-called SLOC (**S**uper **L**arge **O**ptical **C**avity) structure is used [15, 22], where the asymmetry between p- and n- layers, i.e. the thickness and the aluminum content of cladding- and waveguide layers, is changed ('EDASLOC' and 'ASLOC' [30], see section 2.1.3 and 3.2 for details).

BALs are edge-emitting diode lasers, i.e. the laser light is extracted perpendicular to the direction of epitaxial growth through the cleaved facets that form the laser resonator. The facets are passivated and coated with several dielectric mirror layers to achieve high reflectivity ($R_r \approx 96\%$) at the rear side and low reflectivity ($R_f \approx 4\%$) at the front side of the emitter. This one-sided emission facilitates the characterization and the integration of multiple emitters into beam combining systems.

In general, the chip geometry varies in chip width $400\,\mu m < w_{chip} < 1000\,\mu m$ and resonator length $1\,mm < L < 6\,mm$. The chip thickness is $\approx 130\,\mu m$, where the epitaxial grown layers contribute $\approx 5\,\mu m$ and the rest is thinned substrate material. The current is injected via a stripe-shaped top contact on the p-side with a typical stripe width of $50\,\mu m < w < 200\,\mu m$, which is an aperture up to two orders of magnitude wider than in ridge waveguide lasers ('RWL', $w \approx 3\text{-}5\,\mu m$,[19, 21]). The emission wavelength can be tuned in the range $600\,nm < \lambda < 1200\,nm$ by adjusting the compositions of the quantum well, which is often grown in the $In_xGa_{1-x}As_yP_{1-y}$ quaternary system.

Nowadays, modern high power broad area lasers with $w = 100\,\mu\mathrm{m}$ deliver $P_{opt} > 20\,\mathrm{W}$ optical output [31] and their peak conversion efficiency η_c exceeds 70% [32, 33]. However, high power operation also implies high operating current I, high power density at the facets and a considerable thermal load. In order to meet these requirements, the design of the BAL devices is adapted as follows.

The widened contact itself brings several advantages, such as reduced series resistance R_s, which improves $\eta_c = P_{opt}/U \cdot I \approx P_{opt}/\left[(U_0 + R_s \cdot I) \cdot I\right]$. In general, a high conversion efficiency is the best measure to guarantee high output- and low thermal power $P_{th} = (U \cdot I - P_{opt})$. However, the heat that remains must be removed and fortunately a widened stripe also allows fast heat extraction out of the laser chip, due to its reduced thermal resistance $R_{th} = \Delta T_{AZ}/P_{th} = \Delta T_{AZ}/(U \cdot I - P_{opt}) \propto (L \cdot w)^{-1}$ [19], where ΔT_{AZ} is the active zone temperature increase. In addition, the emitters resonator length L is increased as far as internal round-trip losses and the soldering process allow in order to further decrease R_s and R_{th}. Furthermore, the threshold current density $j_{thr} = I_{thr}/(L \cdot w)$ is reduced in stripes with extended area due to reduced lateral current spreading, which lowers the risk of a bulk damage at increased bias. As a further measure, the contact stripe is retreated from the emitter edges in order to reduce front facet heating [3]. In contrast to the BAL in figure 2.1, the contact opening in the BALs of this study is fabricated via He^+ implantation of the p-cap layer. By that measure, a planar emitter surface is produced which facilitates heat extraction and avoids strain fields from mesa etching.

Mathematical description of BAL performance

There is a set of equations that physically describe a diode laser. This set encompasses the rate equations for carrier injection and recombination, the laser threshold condition in Fabry-Perot type edge emitters, a logarithmic gain model and further useful expressions that allow the determination of internal laser parameters, i.e. the internal efficiency η_{int}, the modal gain Γg_0 and the transparency current density j_{tr}. However, these equations are thoroughly derived and explained in detail in the above mentioned literature [19, 20] and in the author's previous thesis [34] (masters thesis, in german language). Hence, only the resulting formulas that parameterize diode laser performance are repeated here for convenience.

In figure 2.2 a typical characterization measurement is shown.

The optical output power P_{opt} above threshold as function of current I, in the regime before power saturation starts, is described via a linear dependency:

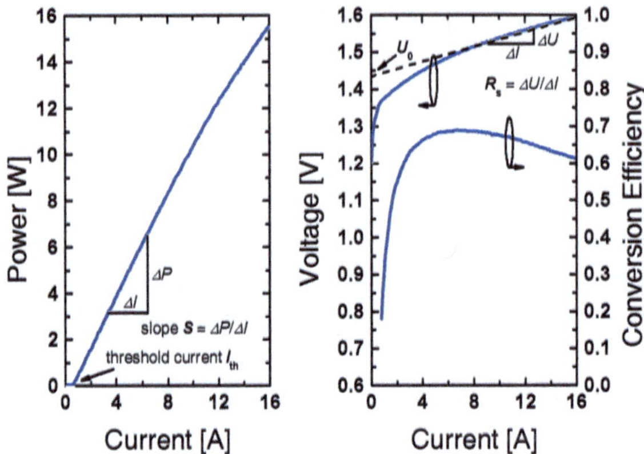

Figure 2.2: Exemplary power- and voltage measurement of a BAL as function of operating current. Left: optical output power P_{opt} increases linearly with current to a slope S as soon as the threshold current value is reached. Right: The diode voltage U follows a curve $I \propto (\exp(qU/k_B T) - 1)$ at low injection, according to the Shockley equation [15, 35] and can be fitted with a linear slope for increased currents to obtain the series resistance R_s and the bandgap voltage U_0 (voltage axis intercept).

$$P_{opt}(I) = \eta_{int} \cdot \frac{h\nu}{e} \cdot \frac{\alpha_{mir}}{\alpha_{mir} + \alpha_{int}} \cdot (I - I_{thr}). \qquad (2.1)$$

Here η_{int} is the internal quantum efficiency, representing the fraction of injected carriers that recombine radiatively in the active zone into useful (lasing mode) photons. The photon energy $E = h\nu$ is the product of Planck's constant h and the optical frequency $\nu = c/\lambda$. The losses are separated into distributed mirror (i.e. out-coupling) losses $\alpha_{mir} = 1/2L \cdot \ln((R_f R_r)^{-1})$ and internal losses α_{int}. The latter term encompasses photon loss mechanisms such as photon absorption on free carriers, absorption in materials outside the active region and/or scattering at rough surfaces. Finally, the threshold current I_{thr} is the operating point where the optical gain compensates all losses such that lasing starts and the output power starts to rapidly increase.

From the slope $S = dP_{opt}/dI$, the external differential quantum efficiency η_{ext} is derived as follows:

$$\eta_{ext} = \frac{S}{h\nu/e} = \eta_{int} \cdot \frac{\alpha_{mir}}{\alpha_{mir} + \alpha_{int}}. \qquad (2.2)$$

The power curve in figure 2.2 starts to bend at $I > 12\,\mathrm{A}$ and the slope decreases. This is the beginning of a so-called 'thermal rollover' that finally leads to a power saturation, mostly due to high chip temperature, that leads to degraded performance. This mechanism is typically described with an empirical law: With increasing (local) chip temperature T_{AZ} in the active region, the threshold current density j and the external efficiency η_{ext} deteriorate according to the exponential functions [22]:

$$j(T) = j(T_r)\exp\left(\frac{T_{AZ} - T_r}{T_0}\right). \tag{2.3}$$

$$\eta_{ext}(T) = \eta_{ext}(T_r)\exp\left(\frac{T_r - T_{AZ}}{T_1}\right). \tag{2.4}$$

Here, T_r is a reference temperature (usually the heat sink temperature, which is set to $T_{HS} = 25\,^{\circ}\mathrm{C}$ in this study) while T_0 and T_1 are characteristic laser parameters that determine how fast the threshold and slope efficiency change with temperature. Typical values for NIR-emitting quantum well BALs are $80\,\mathrm{K} < T_0 < 150\,\mathrm{K}$ and $200\,\mathrm{K} < T_1 < 800\,\mathrm{K}$.

As soon as T_0 and T_1 are known for a given vertical structure, the expected power curve can be emulated more accurately, including the thermal rollover [22]. Here, the voltage curve is approximated as a linear function $U = U_0 + R_s I$:

$$P_{opt}(I) = \eta_{ext}\exp\left(-\frac{R_{th}\left[I(U_0 + R_s I) - P_{opt}\right]}{T_1}\right)$$
$$\times \left[I - I_{thr}\exp\left(\frac{R_{th}\left[I(U_0 + R_s I) - P_{opt}\right]}{T_0}\right)\right]. \tag{2.5}$$

Equation 2.5 can be solved iteratively for incremental current steps by choosing a convergence radius ΔP_{opt} for the output power values. Finally, equation 2.1 is used to derive a formula for the conversion efficiency η_c:

$$\eta_c = \frac{P_{opt}}{IU} = \eta_{int} \cdot \frac{\alpha_{mir}}{\alpha_{mir} + \alpha_{int}} \cdot \frac{h\nu}{e \cdot (U_0 + IR_s)} \cdot \frac{I - I_{thr}}{I}. \tag{2.6}$$

Equation 2.6 illustrates the necessary measures for η_c improvement: enhancement of η_{int} and reduction of α_{int}, R_s and I_{thr}. Moreover, as shown in equation 2.5, an increase of the temperature degradation parameters T_0 and T_1 is helpful, leading to increased values for the saturated power P_{sat}. However, the difficulty lies in improving one of these factors without compromising one of the others.

2.1.3 Epitaxial design of modern BALs

The vertical sequence (i.e. in the direction of epitaxial growth) of semiconductor layers in a diode laser is the most basic aspect of its design as it determines the vertical mode structure, the heat extraction and the efficiency. It requires exact knowledge of the material properties in use and a careful analysis of the resulting band structure, the carrier transport and the optical properties. Here, all laser structures are designed in the $Al_xGa_{1-x}As$ material system, which is well established and manufacturable in high quality.

In general, the design of the vertical layer structure for BALs follows some simple guidelines:

1. construct a waveguide such that the fundamental vertical mode dominates and higher order modes are suppressed

2. increase the mode expansion for reduced facet load and increased power output

3. optimize thickness, refractive index and doping-level of all layers such that:

   - waveguide losses are minimized
   - carriers are well confined
   - electrical resistance is low
   - sufficient gain is provided
   - efficiency is enhanced

4. the design needs to be robust, i.e. slight thickness deviations in the layers do not change the BAL performance significantly

Depending on the intended application of the lasers other criteria might be added, e.g. low vertical far field divergence, sufficient p-side thickness for wavelength-stabilization via surface gratings or optimized material consumption for reduced costs.

The study presented in this work profits from a process of continuous design improvements at the FBH, resulting in very efficient vertical laser structures with $\eta_c > 65\%$ at $P_{opt} = 10\,W$ [33]. As shown in the tables 1.2 and 1.3 these structures can compete with highest industry standards and recent research developments. The details of the vertical designs that were used in this study will be presented in the following paragraphs.

As mentioned above, the structures in this work are based on the 'SLOC' concept. This structure widens the optical mode for decreased power density at the front facet. The use of high quality strained quantum wells assures high material gain and carrier confinement, such that the mode overlap (vertical confinement factor Γ_x) and the doping near the active region can be lowered. The latter reduces the internal losses caused by free carrier absorption. However, the doping is increased in regions that are not penetrated by the laser mode to assure carrier injection into the active area and a low series resistance. In particular, the p-cap layer of the diode is highly doped to create an ohmic contact that does not hinder the current flow.

Asymmetry in vertical laser structures

In asymmetric vertical structures, the refractive index and thickness of p- and n-side of the waveguide and the cladding are varied [30, 36, 37]. Since the carrier mobility in p-doped GaAs is at least one order of magnitude lower compared to n-doped GaAs [15], the goal is to reduce the p-side thickness for reduced electrical resistance. Furthermore, a thinned p-side reduces bias induced carrier leakage [30]. In addition, since the density of states is higher in p-type GaAs (increased effective mass for holes), the optical losses due to free carrier absorption are increased as well [38]. Hence, the fundamental mode is constructed with minimized overlap in the p-side to reduce internal losses.

In table 2.1 and figure 2.3, the two main designs for this study are introduced. Their performance will be analyzed (cf. section 3.2) with respect to their different 'degree' of asymmetry, which is supposed to influence the lateral beam quality.

layer	ASLOC	EDAS
p-cladding	$0.4\,\mu\text{m}$ $\text{Al}_{0.85}\text{Ga}_{0.15}\text{As}$	$0.6\,\mu\text{m}$ $\text{Al}_{0.85}\text{Ga}_{0.15}\text{As}$
p-waveguide	$0.73\,\mu\text{m}$ $\text{Al}_{0.15}\text{Ga}_{0.85}\text{As}$	2 step GRIN $0.22\,\mu\text{m}$ $\text{Al}_x\text{Ga}_{1-x}\text{As}$: $0.85 \geq x \geq 0.2$
active zone	SQW InGaAs, $\lambda = 962\,\text{nm}$	DQW, InGaAs, $\lambda = 910\,\text{nm}$
n-waveguide	$0.83\,\mu\text{m}$ $\text{Al}_{0.15}\text{Ga}_{0.85}\text{As}$	3 step GRIN $2.6\,\mu\text{m}$ $\text{Al}_x\text{Ga}_{1-x}\text{As}$: $0.2 \geq x \geq 0.35$
n-cladding	$2.1\,\mu\text{m}$ $\text{Al}_{0.19}\text{Ga}_{0.81}\text{As}$	$1.5\,\mu\text{m}$ $\text{Al}_{0.35}\text{Ga}_{0.65}\text{As}$

Table 2.1: Structure details of MOVPE-grown vertical designs in test.

The first design is an asymmetric super large optical cavity (ASLOC) with a 0.73 μm thick p-waveguide and asymmetric aluminum content in the n- and p-cladding. As can be seen in figure 2.3(a), the fundamental mode expands symmetrically into the n- and p-waveguides. In contrast to that, the second design in figure 2.3(b) (EDASLOC, [30]) is designed to have minimal overlap in the p-part of the device. A thinned p-waveguide (0.22 μm) and an expanded GRIN n-waveguide cause the fundamental mode to expand widely into the n-region instead. This shift also causes reduced optical confinement in the EDASLOC design, which is compensated by using a double quantum-well (DQW) in the active region.

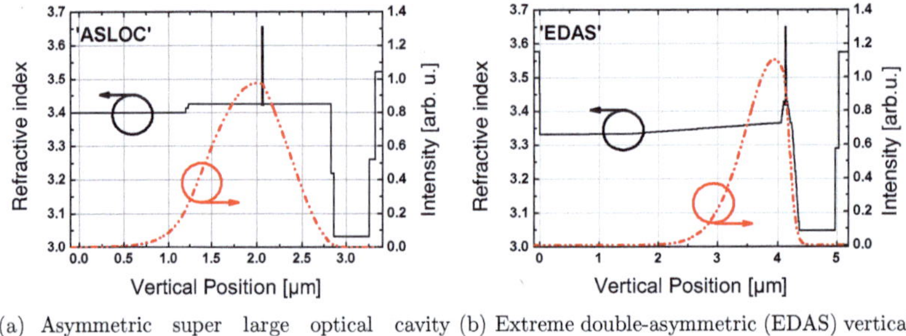

(a) Asymmetric super large optical cavity (ASLOC).

(b) Extreme double-asymmetric (EDAS) vertical design.

Figure 2.3: Vertical profiles of refractive index (solid, black) and intensity of fundamental mode (red, dash-dotted) for the two epitaxial designs in test.

2.1.4 Beam quality and radiance of a BAL

The properties of modern BALs that were examined so far are very promising and make them suitable for applications in materials processing: High power output, high efficiency, compactness, wavelength tunability and low cost mass production make them the ideal device for power scaling in beam combining systems.

However, for use in these systems, the laser *beam* needs to fulfill two important criteria. First, high polarization purity is necessary to enable polarization multiplexing [3]. Secondly, the laser beam must be focusable to very small spots to achieve sufficient power density for e.g. sheet metal cutting and to enable coupling into low-NA optical fibers. ('NA' = numerical aperture, describes the fiber acceptance angle α for known refractive index n_m of a surrounding medium: NA $= n_m \cdot \sin \alpha$.)

In single emitter broad area lasers, the first criterion is fulfilled sufficiently. The use of compressively strained quantum wells separates the valence bands for light- and heavy holes such that only electron to heavy hole recombination occurs, which leads to in-plane (TE-) polarization of the laser emission [22]. However, as will be shown in section 3.3.2, fabrication-, and packaging induced strain fields can distort the polarization purity. Nevertheless the degree of polarization (DOP) is very high in modern single emitter BALs $\rho' = P_{TE}/P_{TE} + P_{TM} > 95\%$. In laser bars, the situation can be very different. Here the $\approx 1\,\text{cm}$ wide chips suffer from strain fields that occur during the soldering process. Usually the DOP varies between the emitters in the bar center and at the edge.

It turns out that the second criterion - focusability - is a far greater problem and this thesis is devoted to it. In order to quantify 'how good a laser beam can be focused', the *beam parameter product* is introduced as figure of merit: Figure 2.4 shows a two-dimensional projection of a laser beam. The two important parameters are the beam width w at the beam's smallest extent (the beam *waist*) and the divergence angle Θ. A common way to measure width and angle is to determine the FWHM (**F**ull **W**idth at **H**alf **M**aximum) from measured intensity profiles. However, in order to adapt to industry standard, these parameters are measured such that 95% of the beam's power is included in them.

The beam parameter product BPP is then defined as:

$$BPP = \frac{w_{95\%} \cdot \Theta_{95\%}}{4}.$$

(2.7)

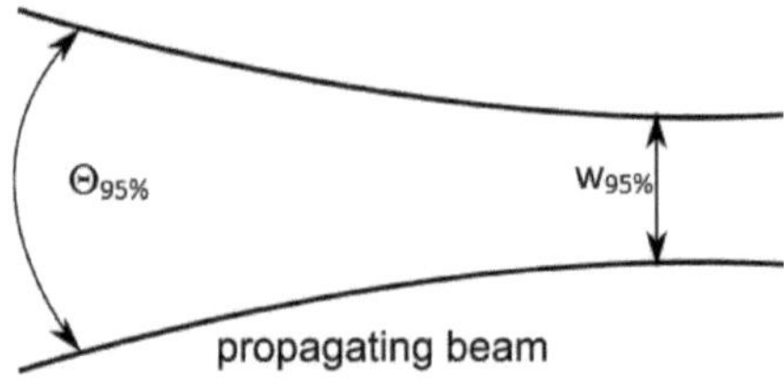

Figure 2.4: Illustration of beam properties: Full divergence angle $\Theta_{95\%}$ and beam waist width $w_{95\%}$.

Please note that this definition uses the *full* angle and waist, which makes the division by 4 necessary. The inverse of the BPP is called *beam quality* $K = 1/BPP$, but it is common practice to directly refer to BPP as 'beam quality' having in mind that the quality is increased as BPP is reduced. The beam parameter product has a natural limit - the so-called *diffraction limit* λ/π - which is wavelength dependent and the BPP of a single mode, Gaussian shaped and fully coherent laser beam. This leads to an alternative figure of merit for beam quality, the *beam propagation ratio M^2*:

$$M^2 = \frac{BPP}{\lambda/\pi}. \tag{2.8}$$

The propagation ratio relates any beam to a fundamental mode Gaussian beam with the same wavelength and gives the factor '... times diffraction limited'. With BPP and M^2 two equivalent figures are given that will replace the vague term 'focusability'. The lower the BPP of a laser beam, the smaller is the focus spot. Finally, a concrete definition for the power density of a laser beam will be given here. The *radiance B* of a light source is defined as follows:

$$B = \frac{P_{opt}}{A \cdot \Omega}. \tag{2.9}$$

The radiance is the quotient of the optical power P_{opt} and the product of the emitter surface A with the beams solid angle Ω. However, the definition in equation 2.9 is quite cumbersome. For rotationally symmetric beams, e.g. from fiber-, disk-, or slab lasers, the radiance is typically expressed via the beam parameter product BPP [3]:

$$B = \frac{P_{opt}}{\pi^2 \cdot BPP^2}. \tag{2.10}$$

However, for diode lasers with an asymmetric emission profile in x- and y-directions (cf. figure 2.1 and 2.5), it holds that the total BPP_{xy} can be expressed via the BPP in the two transverse beam directions: $BPP_{xy} = \sqrt{BPP_x \cdot BPP_y}$. Hence, the radiance in this case is given by [3]:

$$B = \frac{P_{opt}}{\pi^2 \cdot BPP_x \cdot BPP_y}. \tag{2.11}$$

Having introduced the important figures, a brief analysis of the BAL laser emission will be done here. How is the emission from a BAL shaped? A closer look at the emission cone in figure 2.1 reveals a strong asymmetry. In the x-direction (also called the vertical- or *fast*-axis), the laser field is confined to a few microns due to the epitaxially grown, well defined waveguide structure. This confinement in the near field leads to a large divergence in the far field. In figures 2.5(a) and 2.5(b), exemplary near and far field profiles are shown (calculated). The near field width is only $w_{95\%} = 1.54\,\mu m$ (1 µm FWHM), but the far field divergence reads $\Theta_{95\%} = 46.2°$ (25.4° FWHM). However, far field divergence by itself is not problematic, if the beam quality is high (i.e. $M^2 \approx 1$) and can easily be imaged by the micro-lenses which are used in beam combining systems. The shape of the vertical profiles shows that only the fundamental mode is excited and the beam quality in that direction is very good, with values $1.1 < M_{vert}^2 < 1.5$ and minimal dependence on bias. The deviation from $M^2 = 1$ is due to the mode shape difference from a pure Gaussian profile.

In the y-direction (also called lateral- or *slow*-axis), the near field expands over several tens of microns, depending on the injection stripe width. Figures 2.5(c) and 2.5(d) show the corresponding intensity profiles, taken from exemplary measurements at $P_{opt} = 10\,W$. Here, the near field width is $w_{95\%} = 92\,\mu m$ (95.8 µm FWHM) and the far field angle is $\Theta_{95\%} = 10.8°$ (9.6° FWHM). Both profiles have a top-hat shape with irregular intensity fluctuations, which is typical for multi-mode laser radiation. The beam quality along the slow-axis is much poorer compared to the fast-axis. Furthermore, the profiles have a strong dependence on operating current I, such that the beam quality varies: $10 \leq M_{lat}^2 \leq 16$.

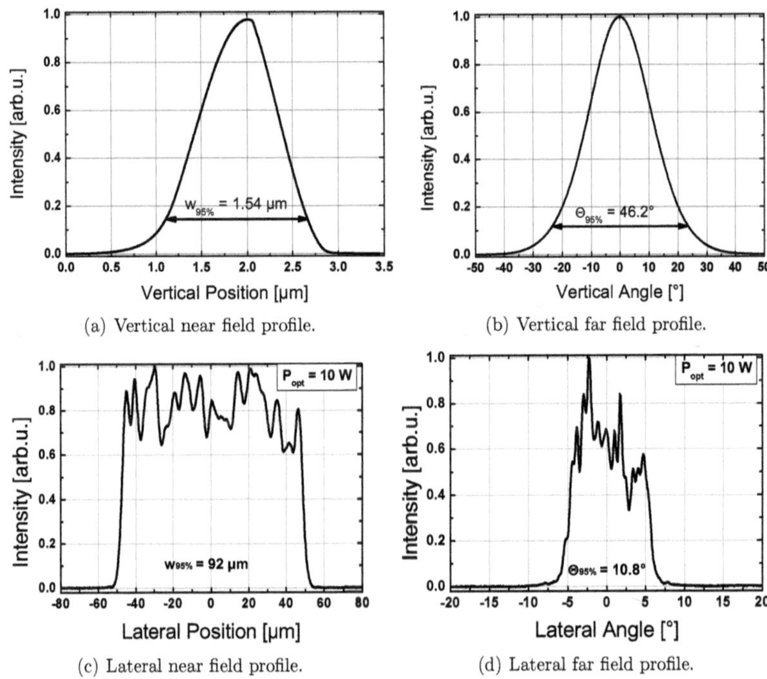

(a) Vertical near field profile.

(b) Vertical far field profile.

(c) Lateral near field profile.

(d) Lateral far field profile.

Figure 2.5: Illustration of broad area laser beam shape ($L = 4\,\mathrm{mm}$, $w = 90\,\mathrm{\mu m}$, $\lambda \approx 910\,\mathrm{nm}$). The emission is separable into two main directions. Vertical near field (a) and (b) far field are single moded, but strong near field confinement causes large far field divergence. Lateral near field (c) and far field (d) are of irregular, multi-moded shape. Large near field extension causes low far field divergence.

Since the beam quality in the vertical direction is close to the diffraction limit and stable for increased currents, the BALs radiance is solely determined by the *linear radiance* $B_{lin} = P_{opt}/BPP_{lat}$. Figure 2.6 shows an exemplary measurement of near- and far field widths at increasing output power, revealing increasing far field divergence and shrinking near field width at increased bias. As a result, the BPP_{lat} increases as well and this growth compensates the growth in output power, such that B_{lin} saturates.

How can the difference in beam quality between the fast and slow axes be explained? The theoretical considerations on M^2 in [40] show that in general the squared beam propagation ratio of an arbitrary beam can be decomposed into three influences:

$$M_i^2 = \sqrt{M_{i,diff}^2 + M_{i,ab}^2 + M_{i,cohe}^2}.$$
(2.12)

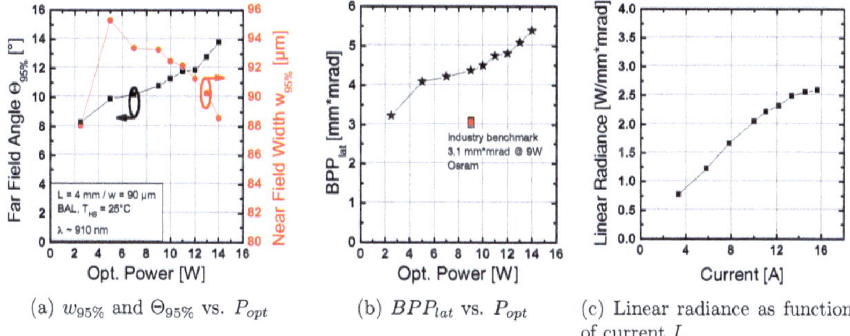

(a) $w_{95\%}$ and $\Theta_{95\%}$ vs. P_{opt} (b) BPP_{lat} vs. P_{opt} (c) Linear radiance as function of current I.

Figure 2.6: Properties of the BAL laser field in lateral direction. (a) With increasing power P_{opt} the far field divergence increases and the near field width shrinks. (b) As a consequence the lateral beam parameter product BPP_{lat} increases with output power. A benchmark value from industry fabricated minibars is indicated (taken from [39]). (c) The linear radiance B_{lin} saturates at high currents due to increased BPP_{lat}.

Here the index i can be substituted by either x or y in order to describe vertical or lateral beam propagation, respectively. The first term $M^2_{i,diff}$ is a measure of the influence of diffraction (due to the steepness of the power density distribution) on M^2_i. The second term $M^2_{i,ab}$ describes the influence of wavefront aberration and finally $M^2_{i,cohe}$ is associated with the lack of coherence of the beam e.g. due to an incoherent superposition of transverse modes. It should be noted, though, that equation 2.12 gives a description of optical beam non-ideality, but cannot provide information on the physical processes behind this.

Now, looking back at the field profiles in figure 2.5, an explanation for the M^2 discrepancy between fast- and slow axis emerges: The field profile in vertical direction consists only of the fundamental mode and shows no intensity fluctuations. Hence $M^2_{x,cohe}$ and $M^2_{x,diff}$ and therefore M^2_x are very small. In the lateral direction, however, the laser field is a superposition of many ($\approx$ 30-40) modes and shows considerable intensity fluctuations, such that $M^2_{y,cohe}$ and $M^2_{y,diff}$ are increased.

There are two main causes for the multi-moded lateral laser field. First, with increasing current, a thermal waveguide builds up that guides multiple lateral modes. The number depends on the injection stripe width, as will be shown in section 4.1. Secondly, the spatially broad gain region (stripe width + a few micron of current spreading) allows the amplification of all guided modes, such that no intrinsic mode discrimination is present. Furthermore, spatially extended gain regions are prone to filamentation [41] that increases $M^2_{y,diff}$ by the formation of steep, peak-shaped filaments in the laser near field (cf. section 3.2.3).

Unfortunately, the large number of lateral modes is not the only problem. Recent investigations [42, 43] show, that the shape of the individual modes (including the fundamental mode) is distorted at high power operation. This observation indicates a further, yet unknown, mechanism that deteriorates BPP_{lat} in a systematic manner.

Since the degradation of the slow-axis beam quality is a major problem in the BAL fabricating industry, measures to suppress it are sought. Studying potentially helpful design measures and determining the effects that are most influential on BPP_{lat} is the core of this thesis that will be presented in chapter 3.

2.2 Toolbox for the optoelectronic device engineer

The device engineers 'toolbox' contains many methods of different type, including empirical relations for material parameters, mathematical solving algorithms, complex simulation tools, processing techniques and characterization procedures. This section contains a brief overview on the methods that are used within this study.

2.2.1 The refractive index of AlGaAs

For the simulation of light propagation in a medium, the refractive index n is most important. Here, the medium is the $Al_xGa_{1-x}As$ material system. An ideal refractive index function encompasses all parameters that have an impact. For a compound semiconductor, this function $n(\lambda, x, N^-, P^+, T)$ depends on the wavelength λ, the material composition x, the concentrations of free electrons N^- and holes P^+ and the temperature T. However, such a function is not available for the whole parameter space and hence a simplified model with a correction term is used:

$$n_{tot}(\lambda, x, N^-, P^+, T) = n(\lambda, x, T) + \Delta n(\lambda, x, N^-, P^+) \tag{2.13}$$

In equation 2.13, the total refractive index n_{tot} consists of a basic value $n(\lambda, x, T)$ and a correction that takes carrier concentration into account [44, 45]. However, in this work simple one dimensional waveguide calculations are done, which focus on the formation of a thermal lens in the active area. The correction due to injected carriers has been neglected here, justified by previous observations [42] that confirm the minimal influence of the carrier profile on lateral waveguiding in modern BALs. However, as shown in [7] and section 3.4, the carrier profile - especially along the device edges - is decisive for the gain supply of higher order modes.

As discussed in [34], there are several models that yield a refractive index function of the form $n(\lambda, x)$. The temperature dependence is then often added with help of a correction term of the form $\Delta n(T) = T \cdot \partial n / \partial T$, assuming a linear temperature response with the slope $\partial n / \partial T \approx$ 4E-4 K^{-1} [46].

In this work, the model from Gehrsitz *et al.* [47] is used since it directly includes the dependency on temperature $n(\lambda, x, T)$ and reproduces the experimental data on $Al_x Ga_{1-x} As$ compounds very well [34]. It was implemented as a MatLab script (see appendix A for details), such that one dimensional temperature profiles can be converted into refractive index profiles.

It should be noted here that, in addition to the above mentioned influences, the material strain also has an impact on the refractive index [48, 49]. That aspect will be discussed in section 3.3.

2.2.2   Calculating lateral modes in BALs

For the analysis of the lateral beam properties, knowledge of the number and the shape of the lateral modes is important. More modes imply poorer beam quality, due to the reduced degree of mutual coherence (see eq. 2.12). Furthermore, the shape of each mode is of interest, since a mode deformation for low order modes would reduce the effectiveness of mode filtering [42]. Here, the focus is set on how the thermal lens shape influences the mode spectrum. In order to assess this influence, a simple one-dimensional mode-solver is used that calculates the guided modes of a lateral thermal waveguide:

The electric field amplitude $\vec{E}(x, y, z)$ in a dielectric medium with refractive index $n(x, y, z)$ can be obtained by solving the time independent Helmholtz equation [22]:

$$\left(\frac{\partial^2}{\partial x^2} + \frac{\partial^2}{\partial y^2} + \frac{\partial^2}{\partial z^2} \right) \vec{E}(x, y, z) + k_0^2 n^2(x, y, z) \cdot \vec{E}(x, y, z) = 0 \tag{2.14}$$

Equation 2.14 is a second order elliptic partial differential equation, where $k_0 = 2\pi / \lambda_0$ is the vacuum wavenumber. The coordinate-system is aligned according to the axes in figure 2.1. In this thesis, a full solution for $\vec{E}(x, y, z)$ is not necessary, since the focus is on the field-distribution in the lateral direction. Hence, restricting the analysis to the active area and assuming TE-polarization, the field amplitude is only non-zero for the y-component:

$\vec{E}(x, y, z) = \{0, E_y(y, z), 0\}$. The propagation in z-direction is separated from the field in the y-direction via a plane wave ansatz: $E_y(y, z) = E_y(y) \cdot \exp(-in_{eff}k_0 z)$. With this ansatz, the Helmholtz equation becomes one dimensional (the y-subscript has been omitted for convenience):

$$\frac{\partial^2 E(y)}{\partial y^2} + [k_0^2 n(y)^2 - k_0^2 n_{eff}^2]E(y) = 0. \tag{2.15}$$

In equation 2.15, the wavelength λ and the refractive index profile in lateral direction $n(y)$ are the input parameters. The solution is obtained with a finite difference approach. Therefore, a discretization with step-size Δy is applied along the y-direction, such that $N_y \cdot \Delta y = w_{chip}$, where N_y is the number of grid-points and w_{chip} is the laser chip width. The electric field is discretized into a column vector $E(y) \rightarrow \vec{E} = E_{\{i\}}$ with $i \in \{1, ..., N_y\}$. The differentiation is approximated by a 3-point finite difference scheme [19]:

$$\frac{\partial^2 E(y)}{\partial y^2} \cong \frac{E_{i+1} - 2E_i + E_{i-1}}{\Delta y^2}. \tag{2.16}$$

Now equation 2.15 reads:

$$\frac{E_{i+1} - 2E_i + E_{i-1}}{k_0^2 \Delta y^2} + n(y)^2 E_i = n_{eff}^2 E_i. \tag{2.17}$$

Equation 2.17 is an eigenvalue problem of the form $\hat{A}\vec{E} = \beta \cdot \vec{E}$, where $\hat{A}$ is a tridiagonal sparse matrix and $\beta = n_{eff}^2$ is the eigenvalue. The set of eigenvectors $\vec{E}_m$ of equation 2.17 represents the lateral modes that are guided by the refractive index profile $n(y)$. The effective refractive index $n_{eff} = \sqrt{\beta}$ of each mode describes its propagation angle within the waveguide.

In this thesis, the problem $\hat{A}\vec{E} = \beta \cdot \vec{E}$ is solved with MatLab's powerful *eigs()*-function that uses the implicitly restarted Arnoldi Method [50]. The field-distributions of all calculated modes are then further processed to far field distributions (using Fourier transformation). In addition, the lateral confinement $\Gamma_{y,m}$ and the lateral beam parameter product $BPP_{y,m}$ of each mode are calculated.

In figure 2.7, an example calculation is shown. The input refractive index profile was converted from a $w = 90\,\mu m$ BAL FEM-temperature profile using $n(\lambda, x, T)$ from [47]. The fundamental mode has the highest n_{eff} and is depicted in figure 2.7(b) together with the 5th and the 10th lateral modes, which show a symmetric, oscillating near field intensity with strong outside lobes. The derived BPP values and the lateral confinement (assuming rectangular shaped, $90\,\mu m$ broad gain profile) are shown in figure 2.7(c). With increasing mode number, the BPP increases but the modal gain decreases since the confinement drops rapidly.

The simple one-dimensional simulation presented here is well suited to *qualitatively* assess the impact of different thermal waveguides and lateral effective index steps. However, a direct comparison of the simulation results with experimental data is difficult, since the measured BPP_{lat} encompasses all modes and there is no clear information on how the power is distributed among the mode spectrum. In addition, a free running BAL also has multiple longitudinal modes that supply further families of lateral modes [42]. Finally, the lateral carrier distribution, filamentation processes and other non-linearities also influence the lateral mode profiles, but are not included in this simple model.

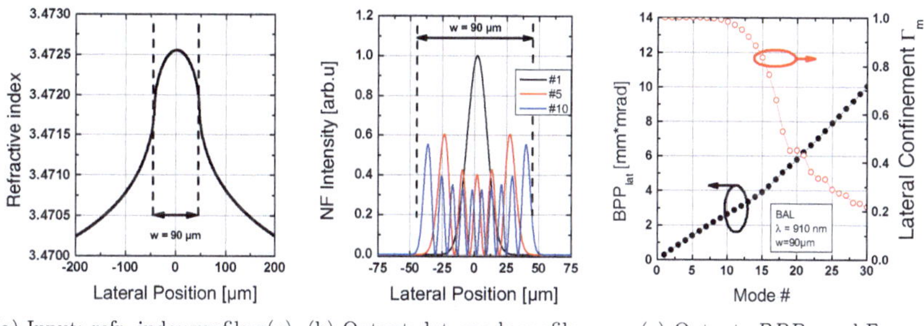

(a) Input: refr. index profile $n(y)$. (b) Output: lat. mode profiles. (c) Output: BPP_m and Γ_m

Figure 2.7: Calculation of lateral modes in a thermal waveguide. (a) The refractive index profile $n(y)$ is given to the simulation procedure as an input parameter. (b) Near field intensity $I_{NF,m} = |\vec{E}_m|^2$ of selected lateral modes. The dashed lines indicate the current aperture $w = 90\,\mu m$, which is assumed to be a uniform gain region wherein the modes are amplified. (c) Resulting $BPP_{lat,m}$ and confinement Γ_m for each mode (assuming separable confinement $\Gamma = \Gamma_{vert} \times \Gamma_{lat}$).

2.2.3 Ion implantation and conductivity of AlGaAs

The purposeful manipulation of the electrical properties of semiconductor material is an important tool for device engineers. There are two main techniques to regulate the electrical conductivity of semiconductors. First, the introduction of dopants during the crystal growth enables a wide range of carrier-concentrations for every layer (either p-

or n-type) of the vertical structure. However, the doping is distributed uniformly inside the semiconductor layer and cannot be changed as soon as the growth is finished. The second method is called *ion implantation*, where the dopant species is ionized and then accelerated into the semiconductor crystal by an electric field. Since the incoming ions can easily be shielded with metal-masks, this technique allows local control of doping type and carrier concentration.

In this work, however, the focus is laid on ion implantation that locally reduces the carrier-density and mobility of doped GaAs layers [51]. Therefore, the implantation is done with light ions, such as protons (H^+) or helium (He^+) in order to reach penetration depths of several microns at acceleration energies of a few hundred keV.

The elastic collision of these light ions with the lattice atoms causes perturbations of the crystal structure in the form of point-defects (anti-site defects and vacancies). These defects form deep-level centers in the middle of the band gap that reduce the number of free carriers (both electrons and holes) via carrier-trapping [52]. These levels are not thermally ionized at temperatures $\leq 600\,°C$, so that the carrier traps remain active during the device operation at room temperature. In addition, the point-defects form scattering centers that reduce the carrier mobility μ_{el} [51]. As a consequence, the electrical conductivity $\sigma_{el} = e \cdot \mu_{el} \cdot N^-$ is reduced significantly in the implanted regions.

The reduction in σ_{el} and the penetration depth are determined by the implanted element, the dose Φ and the acceleration energy E_{impl}. In general, implantation elements with increased mass produce more damage to the crystal and create more carrier traps per incoming ion [52]. On the other hand, the cross section with the lattice atoms increases as well [51], such that the penetration depth (at constant E_{impl}) decreases. The total damage is regulated by the dose Φ, which typically ranges between 10^{14} and 10^{15} cm^{-2} for proton implantation. The damage profile can be simulated using the Monte-Carlo technique, which is implemented in the SRIM software [53]. In figure 2.8, a series of proton damage profiles is shown for varying acceleration energies.

The protons lose energy to electrons and in elastic collisions with the lattice atom nuclei. In order to assure high crystal damage, the propagation along a crystal symmetry axis (*channeling*) is avoided by tilting the target by a few degrees. As the cross section of the protons with the nuclei of the lattice atoms is proportional to $1/E_{H^+}$ [51], most of the damage is produced in the last third of the proton path-length. After the initial collision with the lattice atoms, the latter can have enough energy to further displace other atoms, which produces a damage-cascade that leaves behind multiple vacancies, Frenkel- and antisite defects [52].

In the fabrication of gallium arsenide based diode lasers, the implantation technique is used to define a contact stripe via the generation of a junction isolation in the p-cap layer [54–56]. This procedure guarantees a defined current aperture and a planar emitter surface. In this work, a shallow helium implantation is used to define the BAL injection stripes.

However, in the studies presented here, implantation is also used to modify the conductivity close to the BAL active region. Here, the crystal defects also cause optical absorption that would increase the internal losses dramatically. In [57], Dyment *et al.* describe how optical absorption in proton implanted p-type GaAs can be annealed without losing the high electrical resistivity. In this work, proton implantation and subsequent annealing was used to modify the current injection and the carrier profile in the active layer in BALs (see section 3.4.1). Therefore, the implantation is placed directly beneath the current path at the emitter edges and ranges down to the quantum well.

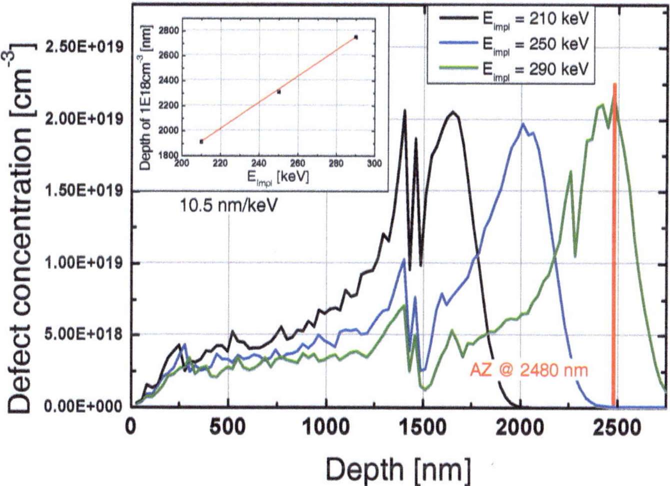

Figure 2.8: SRIM simulation of proton implantation induced damage profile in an exemplary EDASLOC structure. The dose is fixed at $\Phi = 10^{14}$ cm^{-2} and the implantation energy is varied in the range $200\,\text{keV} < E_{H+} < 300\,\text{keV}$. With increasing energy, the damage peak shifts to deeper regions. The inset shows that the depth of a fixed defect concentration (here 1×10^{18} cm^{-3}) increases linearly with E_{H+} at a rate of $10.5\,\text{nm keV}^{-1}$.

Stain etch method for implantation diagnosis

There are multiple analysis techniques to verify successful proton implantation. Therefore, the assessment is divided into multiple steps. The first measure is to resolve whether the implantation profile was realized as intended. Here, a cross-sectional stain etch with subsequent SEM-analysis is very helpful. This technique will be described in the following paragraph. Moreover, cross-sectional and 'top-view' cathodoluminescence (CL) measurements can be done to check if the implantation distance d_{di} was processed correctly. The next question is: If the implantation was successful, did it influence the carrier profile? Here, measurement of the high-angle spontaneous emission allows a quantification of carrier non-clamping in the BAL edges, which is an indirect indication of successful suppression of carrier injection and leakage into the implanted edge region. Moreover, the width of ASE (amplified spontaneous emission) near field profiles is another indicator. The third step is to check the influence of the modified carrier profile on the BAL performance. The latter will be shown in section 3.4.2. The CL-analysis and the high-angle measurement of spontaneous emission will be presented in section 3.4.1.

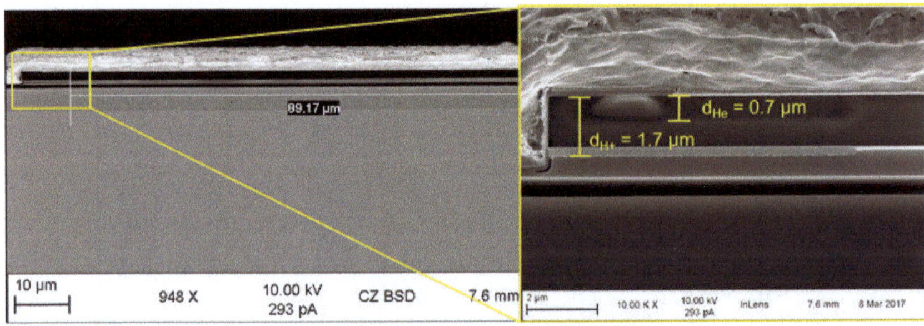

Figure 2.9: SEM images of lateral-vertical cross section of deeply implanted emitters after 5 s of stain etching. Left: Full front view of BAL with $w = 90\,\mu$m. Right: Magnified excerpt of BAL left edge with contact layer Helium implantation and deep proton implantation.

However, in the course of this thesis a stain etch recipe was developed that will be presented here briefly, since it is a useful technique to verify whether the actual implantation depth matches the intended depth. The technique is based on the fact that the etch-rate is increased for an implantation damaged semiconductor. The cleaved BAL facet (cleavage should be done right before the etching) is immersed into the etching solution in order to achieve a delineation of the implanted areas. The recipe for the etching solution originates from [58, 59] and has the following ingredients: 8 g potassium ferrocyanide $K_3\,[Fe(CN)_6]$, 12 g potassium hydroxide KOH and 200 ml water H_2O. However, an initial test with this solution revealed that the etching-time is critical. Even at $t_{etch} = 1\,$s the structures showed extreme over-etching. Hence the above mentioned solution was further diluted with water in the ratio 1:20 in order to be more flexible with the etching-time, where the best result was obtained at $3\,\mathrm{s} < t_{etch} < 7\,\mathrm{s}$.

In figure 2.9, an example of a stain-etched ($t_{etch} = 5\,\mathrm{s}$) cross section of a deeply implanted BAL is shown in an SEM image. The magnified excerpt shows that the contact layer helium implantation reaches down to $0.7\,\mu\mathrm{m}$ and the deep proton implantation stops at a depth of $d_{H+} = 1.7\,\mu\mathrm{m}$.

2.3 Measurement techniques for BALs

In this section, the measurement configurations for the characterization of high power broad area lasers are described. The meticulous characterization of every diode laser under study and the detailed analysis of the measured data are the key to a reliable knowledge enhancement. The measurement techniques need to be precise and repeatable within the error ranges (ideally 10x more precise than the magnitude of the trend that is observed). In the data analysis, it is the task of the scientist to separate artifacts and 'freak' diode lasers from a systematic trend that originates from the intended design changes.

Packaging of BAL single emitters

A processed wafer is structured into test regions, themselves being structured into bars. After fabrication, the bars are cleaved and their facets passivated and coated. One bar consists of 20 single emitters, which can vary in their lateral structuring. Then initial LIV tests are performed on every emitter on a bar in pulsed operation mode up to $2\,\mathrm{A}$ to screen the available material. The most promising devices are then selected for packaging and further testing.Figure 2.10(a) shows the packaging format used in this study. The BALs are soldered with AuSn onto CuW expansion matched carriers. These carriers have 2 contact terminals that are connected to the laser chip via gold bond-wires. For testing in the laboratory, the carriers are clamped with a defined torque to gold-plated copper blocks that are then screwed on the heat sink of the measurement station. The clamp holder (cf. figure 2.10(b)) allows access to both carrier terminals, such that a 4-terminal voltage measurement can be done that eliminates errors due to the voltage drop along the connecting cables, i.e. current is delivered via one contact and voltage measured on the other.

In the following, the different measurement stations will be described. However, they all have in common that the diode laser is thermally stabilized on a gold plated copper heat

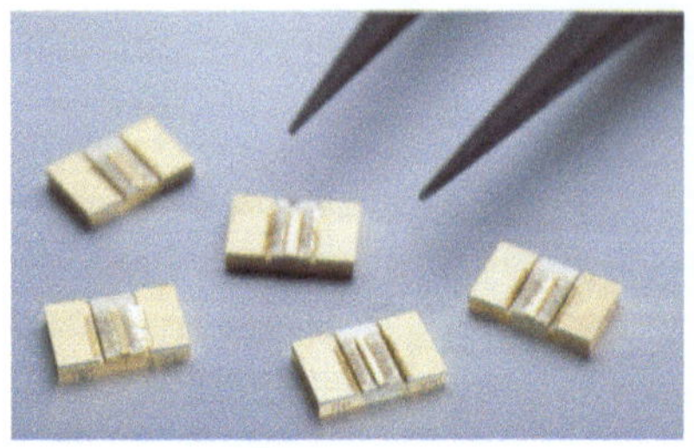

(a) CuW carriers

(b) Clamp holder

Figure 2.10: Standard test format in this study: (a) BALs soldered to CuW carriers for use in two-terminal clamp holders (b).

sink and connected to a diode laser driver. As shown in figure 2.10, the diode laser is soldered to a CuW carrier that is clamped with a defined torque (40 cN m) into a clamp holder. This 'unit' is then used on all measurement stations.

The clamp holder comes with two Pt-100 platinum resistors, one in the holder copper body and one pressed against the rear side of the CuW carrier. If not explicitly stated otherwise, the temperature control is done with the latter sensor. The sensor signal is processed in a thermo-electric cooler (TEC, Newport Model 3150) that regulates the temperature via a Peltier element placed beneath the heat sink plate. The generated heat is then transferred to a water-cooling cycle that is connected to the heat exchanger below the Peltier element. In general, the diode laser characterization is performed at room temperature, such that the heat sink is stabilized on $T_{HS} = 25\,°C$, within a $\pm 0.2\,K$ temperature stability range.

2.3.1 Power current curves and spectra

The most basic evaluation of the BAL performance is a power-current-voltage curve (LIV), where the diode voltage U and optical output power P_{opt} are measured as function of drive current I. Therefore, one terminal of the clamp holder is connected to a current driver (ILX36085-12, accuracy $\pm$ 20 mA) and the other terminal is connected to a multimeter (Keithley Model 2001) in order to measure the diode voltage in 4-terminal configuration (accuracy $\pm$ 50 µV). The output power in continuous wave (cw) operation is measured via a thermopile power detector that is placed in a fixed distance to the laser facet, as shown schematically on the left in figure 2.11. The detector (Gentec type UP19K-15S-W5) is calibrated annually with international standards (ISO 11554). It allows a maximum power of $P_{opt} = 15$ W within a measurement uncertainty of $\pm$ 2.5% and is always used at a fixed spatial offset from the BALs to guarantee consistent illumination of the detector surface [33].

The standard current range for BAL characterization is 0-15 A that is swept in $\Delta I = 100$ mA steps. After acquiring the voltage and output power, the measurement software automatically calculates the conversion efficiency $\eta_c = P_{opt}/I \cdot U$ for every current step.

For the acquisition of spatially integrated spectral mappings, the thermopile detector is replaced by an integrating sphere (cf. figure 2.11 on the right) that collects all the laser radiation and disperses it internally, so that the inner surface of the sphere is illuminated homogeneously. A fraction of the light is then coupled into a spectrometer (Ocean Optics HR4000 operating in the range 890 nm $< \lambda < 1050$ nm with a spectral resolution of 120 pm) via fiber. The fiber coupling terminal can be equipped with additional ND filters in order to reach the full dynamic range and avoid CCD saturation.

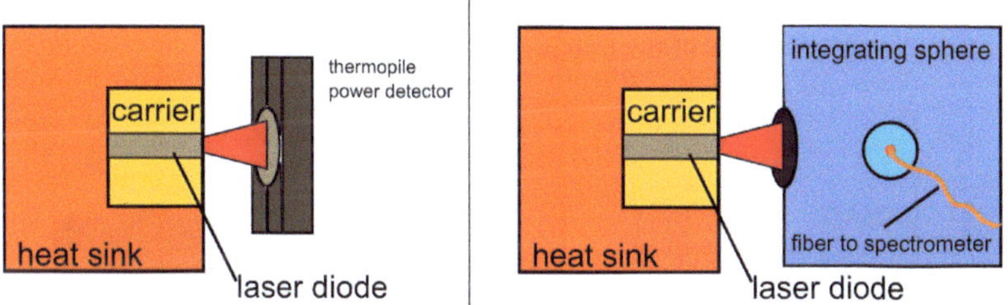

Figure 2.11: Sketch of basic BAL test stations. Left: Measurement setup for LIV characterization. The BAL-carrier is clamped onto a holder that is screwed to a water-cooled heat sink. The output power P_{opt} is measured with a thermopile power detector as function of drive current and in parallel the diode voltage is measured in a 4-terminal configuration. Right: For spectral mappings, the diode laser radiation is collected in an Ulbricht sphere that is connected via fiber to a spectrometer.

2.3.2 Degree of Polarization (DOP)

In broad area lasers with strained quantum wells, as used in this study, the (in-plane) TE-polarization is dominant due to the strain-, and layer thickness-induced separation of the light-hole and heavy-hole valence bands [22]. However, 100% polarization purity is not guaranteed, since for example pressure and strain induced by the diode laser packaging and fabrication can induce local shifts of the electric field polarization. In this thesis, the degree of polarization is defined as follows:

$$\rho' := \frac{P_{TE}}{P_{TE} + P_{TM}}. \tag{2.18}$$

Here P_{TE} and P_{TM} are the output powers measured after a horizontal (TE) and a vertical (TM) orientated polarization analyzer, respectively. Since the diode lasers are used in beam combining systems that also use polarization multiplexing [3], the DOP is critical and limits the maximum combining efficiency. The measurement setup for DOP is depicted in figure 2.12.

Here, the laser emission is first collimated vertically with a fast axis collimator lens ('FAC', cylindrical, plane-convex) close to the facet. Afterwards, the slow axis collimation (SAC, cylindrical, plane-convex) is performed. An aperture cuts off potential stray light and then the beam is directed through a polarizing beam splitter cube (PBS), which is rotated by a stepper motor. The transmitted light intensity is collected in an integrating sphere that is connected to a photo detector. After a 270° rotation of the PBS, the light intensity is plotted as a function of rotating angle and a sinusoidal fit is used to identify the minimum (corresponds to P_{TM}) and the maximum (corresponds to P_{TE}) of the transmission. The total measurement accuracy of the DOP is $\approx \pm 1\%$.

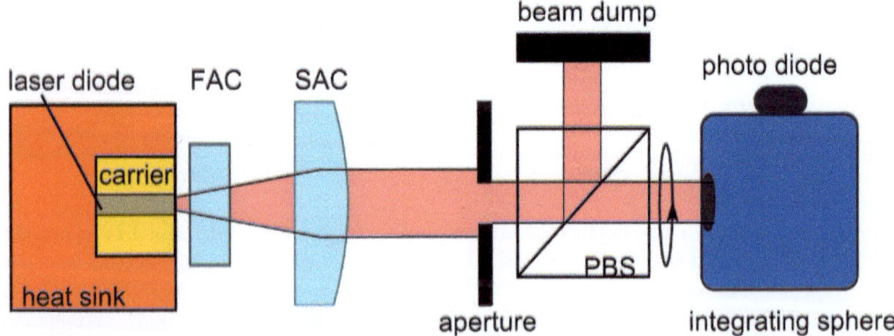

Figure 2.12: Sketch of DOP test station in top view. The laser emission is collimated and analyzed behind a polarizing beam splitter cube, which is rotated by a stepper motor.

2.3.3 Lateral near- and far field profiles

The measurement of the lateral near- and far field intensity distributions and the subsequent calculation of the lateral beam parameter product BPP_{lat} is the key technique in this thesis. The setup (cf. figure 2.13) is built in a moving-slit configuration [60, 61] that allows the acquisition of beam waist (in BALs identical with the front facet near field) and far field profiles with minimal changes in the optical elements.

The diode laser clamp holder is screwed to the heat sink and thermally stabilized as described above. For the near field measurement the lenses 'A' and 'B' form a Kepler-type telescope that magnifies the laser facet by a factor $m_{NF} = f_B/f_A$ into the detector plane. Therefore the distance between the collimating lens 'A' (Thorlabs C240TME-B cyclic aspheric lens, $f_A = 8\,\text{mm}$, NA = 0.5, AR-coating for $600\,\text{nm} < \lambda < 1050\,\text{nm}$) and the imaging lens 'B' (achromatic cyclic lens, $f_B = 500\,\text{mm}$) is fixed at $d_{A-B} = f_A + f_B$ and the distance between lens 'B' and the detector plane is equal to f_B.

The adjustment of the laser beam position can be monitored by directing the beam onto a CCD-camera (Pulnix Model TM-6CN) using a moveable mirror. In the detector plane, two reverse-biased silicon photo diodes (Hamamatsu Model S2386-44K) are placed behind a vertical metal slit (left: $10\,\mu\text{m}$ aperture, right: $50\,\mu\text{m}$ aperture) and the detector output signal is processed in a Keithley multimeter (Model 2001). The Si-diodes are mounted on a pair of linear stepper motors such that the intensity profile in the detector plane can be scanned by both slit types.

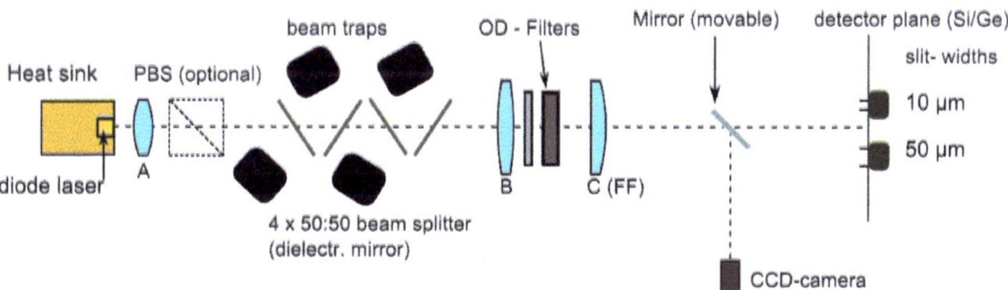

Figure 2.13: Schematic drawing of moving slit setup for near and far field profile acquisition. The lenses 'A' and 'B' form a telescope that magnifies the diode laser front facet into the detector plane. For the acquisition of the far field, a further cylindrical lens 'C' is introduced that images a far field image in the detector plane. For the measurement of polarization resolved profiles a polarizing beam splitter can be introduced behind lens 'A'.

After passing lens 'A', the collimated beam is directed through a cascade of four 50:50 beam splitters (Thorlabs BSW17, 2") in order to protect the CCD-camera and the photo detectors from damage at increased power levels. A further fine-tuning of the beam's total intensity is possible by using a cascade of OD filters behind lens 'B'. For the acquisition of polarization resolved near- and far field profiles, a polarizing beam splitter cube can be placed behind lens 'A' (cf. section 3.3.2). The minimal resolution in the near field profiles is limited by the lasing wavelength $\lambda \cong 1\,\mu m$ and the numerical aperture NA $= 0.5$ according to the Abbe limit $d_{min} = \lambda/\text{NA} \approx 2\,\mu m$.

For the far field measurement, a further lens 'C' (cylindrical lens, $f_C = 400\,mm$) is introduced at a distance f_C to the detector plane. This lens creates an intermediate focus spot and as the focus distance to the detector plane is much larger than the Rayleigh length (z_R: distance from beam waist to the position where the cross-sectional area is doubled, $z_R = \pi w_0^2/\lambda$ for Gaussian beams with beam waist radius w_0), a far field image is created that can be recorded by the photo-detectors. Therefore the detector position x is converted into a far field angle according to the following formula:

$$\theta = \frac{f_B}{f_A \cdot f_C} \cdot x \tag{2.19}$$

Equation 2.19 is valid in the paraxial approximation ($\tan(\theta) \approx \theta \approx \sin(\theta)$, where $\theta \leq 10°$ is the half-angle, such that $\Theta = 2 \cdot \theta$) of laser beam propagation and can be derived using the transfer matrix method of geometric optics (cf. section B in the appendix). The far field angle is measured with an accuracy of $\pm\, 0.1°$.

The setup for near- and far field acquisition described above is used as a standard characterization setup at FBH and throughout this study. Hence, it is used by many different operators in several projects. Therefore, a look on the repeatability is worthwhile. In figure 2.14, a series of 93 measurements on a 'reference' BAL is presented. The tests were distributed over a time span of 3 years and every time the same operating point ($P_{opt} = 5\,W$, $I_{op} = 6.3\,A$) was used.

As shown in figure 2.14(a), the fluctuation in the near field width amounts to $\pm\,2\,\mu$m and in the far field width to $\pm\,0.13°$. The resulting repeatability range for the BPP_{lat} (cf. figure 2.14(b)) is $\pm\,0.1\,$mm mrad. A detailed study of the individual measurements reveals that the deviations from the mean value in the later measurements (starting in 2015) stem from changes in the BAL far field profile itself. Hence, the repeatability is not solely a statement about the measurement setup, but also on the BAL that was used as reference, indicating that the presented values are upper bounds for the setup repeatability. However, other contributing factors for the deviations are the attrition of clamp holder and CuW carrier such that the thermal contact is deteriorated over time and the differing alignment accuracy.

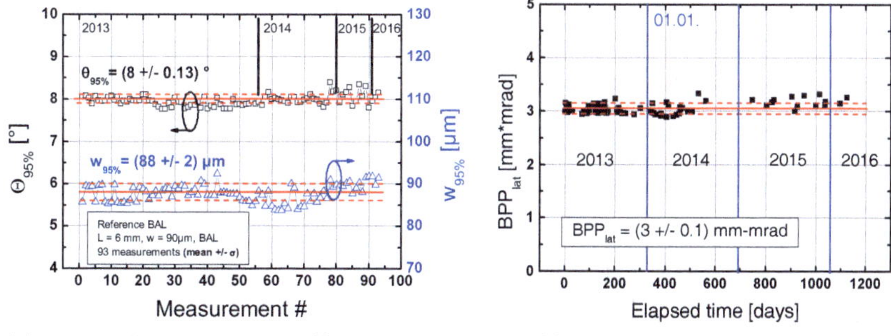

(a) Near and far field widths at 95% power content.

(b) BPP_{lat} long term repeatability

Figure 2.14: Plot of long term repeatability of near- and far field measurement. The same BAL ($w = 90\,\mu$m, $L = 6\,$mm, $\lambda \approx 930\,$nm) was tested at the same operating point ($P_{opt} = 5\,$W, $I_{op} = 6.3\,$A) on 93 different days, beginning in February 2013 and ending in March 2016. Left: Near field width repeatability limited to $\pm\,2\,\mu$m and for the far field width $\pm\,0.13°$. Right: Resulting BPP_{lat} long term repeatability range is $\pm\,0.1\,$mm mrad.

2.3.4 Micro thermography on broad area lasers

In this work, the shape of the lateral-vertical temperature distribution is of great interest, since it determines the thermal waveguide. Hence, a micro-thermography technique, developed at the Max Born Institute [62, 63], was used to determine both the active zone temperature increase and the shape of the thermal lens.

In figure 2.15, a sketch of the thermal camera setup is depicted. Here the BAL is placed on a heat sink in front of the camera objective and a dichroic mirror spatially separates the NIR wavelengths $\lambda < 1\,\mu$m from the mid-infrared thermal radiation. The latter is imaged onto the camera detector chip to yield a 50x magnified image of the laser chip front facet. However, due to the use of a mid-infrared signal, the spatial resolution is limited to no better than $8.8\,\mu$m.

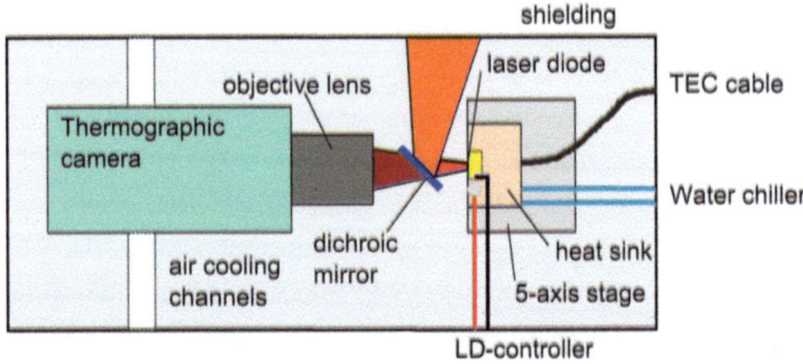

Figure 2.15: Sketch of micro-thermography setup at the Max Born Institute (based on sketch in [62]). The laser diode on carrier is clamped to a holder that is screwed to a heat sink. A Pt-100 thermo-sensor is pressed to the rear side of the carrier, such that the carrier temperature is TEC-controlled during cw operation. The NIR laser emission with $\lambda < 1\,\mu\mathrm{m}$ is separated from the thermo-signal with a dichroic filter between laser diode and thermal camera.

The IR camera signal is superimposed by a slowly varying background, which is eliminated by using differential signals only. The resulting temperature is a spatial average from $\approx 1\,\mathrm{mm}$ information depth, which corresponds to 1/4 of the cavity length, minimizing the influence of overheating effects at the facet, which occur within a few tens of $\mu\mathrm{m}$ [64]. However, the drawback of this large information depth is a lateral broadening effect, that smoothens and widens the recorded thermal profiles (see section 3.2.1). It is caused by the propagation of the infrared radiation inside the semiconductor, mainly determined by the material absorption α and multiple internal reflections [65].

For a correct temperature scale, a calibration procedure is conducted as follows. The temperature controller is connected to a TEC beneath the heat sink and a platinum resistor which is pressed onto the rear side of the carrier. Now, the controller is used to adjust fixed temperature levels of the laser chip and the carrier. By storing camera images for every temperature step, an intensity-to-temperature calibration is possible for every detector pixel. In general, however, a region of interest (ROI) is defined such that only one specific material is contained. Here, the ROI contains only pixels that belong to the GaAs laser chip. Then a third-order polynomial regression is done for every pixel in the ROI and finally all polynomial coefficients are averaged to yield the temperature calibration, which factors in the emissivity of the ROI-material. In order to get a valid calibration curve the maximum temperature level that is set by the controller needs to exceed the maximum junction temperature that is reached later in the cw-mode.

Chapter 3

Main Influences on BAL Slow Axis BPP

In this chapter, a detailed analysis of the effects that influence the slow axis beam quality is presented. According to the literature in this field, the most important factors are thermal blooming [46, 66], the lateral carrier/gain distribution [67, 68], index guides (built-in [69] and unintentional [48]) and filamentation [41, 70]. In figure 3.1, these factors are depicted in a root cause diagram. It also contains mechanical strain and the epitaxial design. The latter influences the beam quality in an indirect manner via its thermal resistance and efficiency, but is also expected to have direct impact via asymmetry-induced mode-control [71].

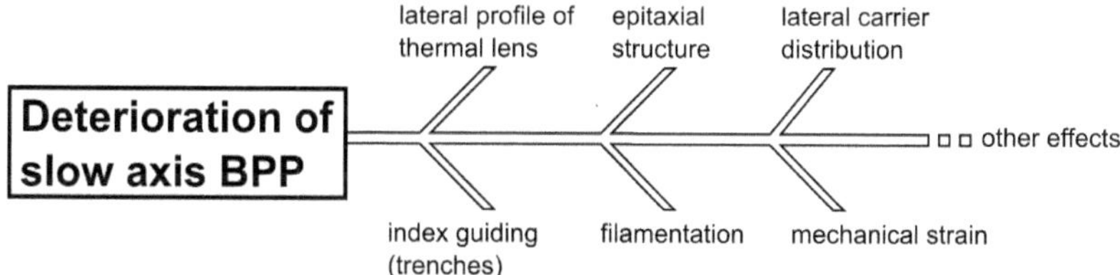

Figure 3.1: List of possible factors for BPP_{lat} deterioration in root cause diagram.

However, there is no detailed information on the relative strength of the listed factors and their interaction. Hence, the goal of this thesis, and specifically of this chapter, is to find a clear picture of the BPP_{lat}-impact of these factors. In order to achieve this goal, an experimental root cause analysis is done. For every potential factor a series of broad area devices is produced that seeks to assess its importance isolated from other effects. Then a comprehensive characterization and data-analysis is performed, including multiple emitters for every BAL species in the test series.

In the following sections, the results of these „isolate-and-investigate"experiments will be presented. First, however, an empirical model will be introduced that enables the comparison of broad area lasers with varying efficiency and thermal resistance.

3.1   Linear model of BPP_{lat} with temperature increase

The BPP_{lat} is usually plotted as a function of output power P_{opt}, which is convenient since this plot also gives information on the radiance. However, in terms of an analysis of BAL properties, this scheme is not sufficient and is inappropriate for a comparative study. The reason for this is that output power describes the eventual result of internal diode laser processes, rather than the processes themselves.

The parameter that is expected to be more meaningful for the internal processes is temperature. It influences the band gap (i.e. the gain spectrum), the electron confinement in the quantum wells and basically all material properties of the semiconductor layers involved, most importantly the refractive index. As outlined in section 2.2.2, the local temperature increase below the injection stripe changes the refractive index of $Al_xGa_{1-x}As$ and therefore induces lateral waveguiding, which directly influences BPP_{lat} via the guiding of multiple lateral modes (thermal 'lensing' or 'blooming').

When two BAL devices are compared at the same output power value, the active zone temperature can differ due to differences in conversion efficiency and thermal resistance. Hence, in this study, the plot of BPP_{lat} as a function of *active zone temperature increase* ΔT_{AZ} is used to provide a better comparability:

$$\Delta T_{AZ} := T_{junct} - T_{HS}. \qquad (3.1)$$

Equation 3.1 defines ΔT_{AZ} as the difference between junction temperature T_{junct} and heat sink temperature T_{HS}. Plotting the BPP_{lat} as a function of ΔT_{AZ} takes into account temperature induced waveguiding, which is determined by *relative* changes in temperature and is not correlated to absolute values. In this scheme, two devices are compared at a state of equal temperature increase of the active zone beneath the injection stripe. Now that the thermal waveguide strength is equal, the differences in BPP_{lat} are caused by other factors, whose relative importance is the subject of this chapter.

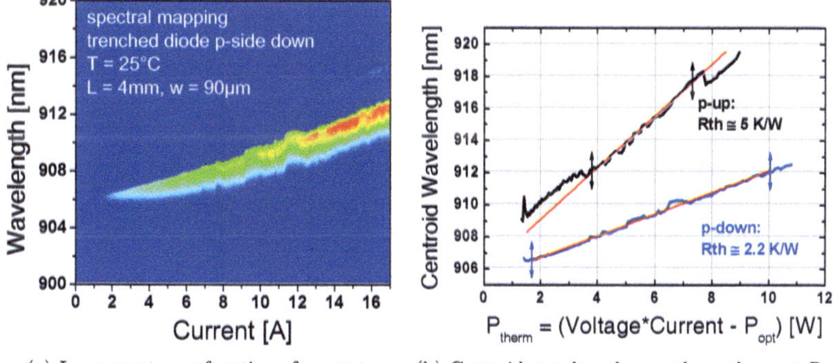

(a) Laser-spectra as function of current (b) Centroid wavelength over thermal power P_{th}

Figure 3.2: Determination of thermal resistance R_{th}. (a) Spectral mappings yield the centroid wavelength λ_c as function of current. (b) A linear regression of the centroid wavelength as function of thermal power gives $\Delta\lambda_c/\Delta P_{th}$, which is used in equation 3.2 to determine R_{th}.

In this study, the active zone temperature increase ΔT_{AZ} was determined in two ways. The first method is to directly measure the BAL lateral-vertical temperature field with a thermal camera, as done in [72]. This technique supplies a great deal of useful information on the internal temperature distribution, which will be presented in section 3.2.1 [1].

The standard method, however, relies on the thermal wavelength drift of the lasing wavelength λ, which is sensitive to the average temperature of the active zone. Here, the determination of thermal resistance R_{th} and thermal (i.e. dissipated) power P_{th} is necessary, since the product of both directly yields the temperature increase $\Delta T_{AZ} = R_{th} \cdot P_{th}$. This method was presented recently in [73] and uses the basic characterization measurements of output power P_{opt}, voltage U and laser-spectra as function of current I. First, a series of spatially integrated spectra is recorded as function of current (cf. figure 3.2(a)), yielding the centroid wavelength λ_c. Afterwards, λ_c is plotted as function of thermal power, which is calculated on basis of the LIV-measurements $P_{th} = P_{opt} - (U \cdot I)$. As depicted in figure 3.2(b), a linear regression is used to determine the shift of λ_c with thermal power $\Delta\lambda_c/\Delta P_{th}$. This shift needs to be multiplied by the reciprocal of the wavelength drift factor $\Delta\lambda_c/\Delta T$ to yield the thermal resistance R_{th}:

$$R_{th} = \frac{\Delta\lambda_c}{\Delta P_{th}} \cdot \left(\frac{\Delta\lambda_c}{\Delta T}\right)^{-1} = \frac{\Delta T}{\Delta P_{th}}. \tag{3.2}$$

---

[1]Special thanks go to Robert Kernke, Martin Hempel and Jens W. Tomm for their support and the provision of a microthermography setup.

The active zone temperature increase can now be calculated for every operating point:

$$\Delta T_{AZ} = P_{th} \cdot R_{th} = (U \cdot I - P_{opt}) \cdot R_{th}. \tag{3.3}$$

In this study, equation 3.3 is used to compare the BPP_{lat} as function of ΔT_{AZ} for BALs with different vertical designs and lateral patterning.

To illustrate that ΔT_{AZ} instead of T_{HS} is the critical parameter for the BPP_{lat}, figure 3.3 shows a series of beam quality measurements on a BAL at five heat sink temperatures between $T_{HS} = 15\,°\mathrm{C}$ and $T_{HS} = 55\,°\mathrm{C}$. Here, independent of the heat sink temperature, the BAL shows the same variation in lateral beam parameter product as function of ΔT_{AZ}, indicating that absolute temperature elevation is *not* relevant for BPP_{lat}. More specifically, this plot implies that the local thermal lens is one of the critical factors in the root cause diagram, as this is expected to be proportional to ΔT_{AZ}.

Before the main influences on the BPP_{lat} are discussed in the subsequent sections, a key observation is noted. In figure 3.3, the BPP_{lat}-functions follow a linear trend with ΔT_{AZ}. It turns out that this is a good approximation for almost all of the broad area lasers investigated here. The evolution of the lateral beam quality with relative temperature elevation is therefore described with a linear model:

$$BPP_{lat}(\Delta T_{AZ}) = BPP_0 + S_{th} \cdot \Delta T_{AZ}. \tag{3.4}$$

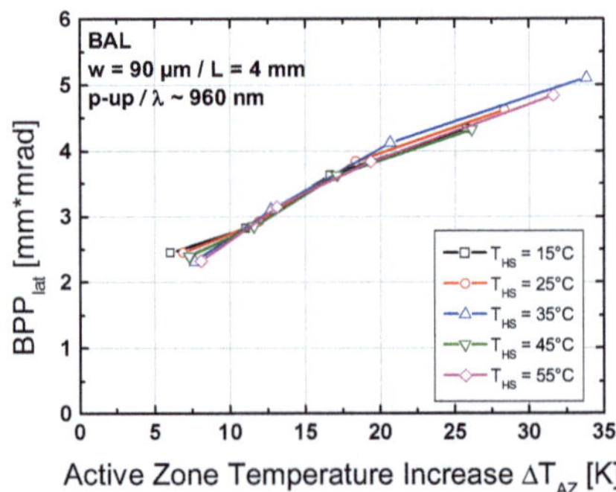

Figure 3.3: Lateral beam parameter product BPP_{lat} as function of ΔT_{AZ} measured on a commercially available BAL at five heat sink temperatures.

Here the *thermal slope* S_{th} represents all mechanisms that impact on BPP_{lat} and are related to the temperature increase. The linear intercept BPP_0 is called BPP-*ground level* and is a measure of all influences that increase the BPP_{lat}, but vary slowly with temperature. In figure 3.4, the linear model is applied to a beam quality measurement of a single emitter broad area laser. However, such a simple approximation is limited in the sense that details in the BPP_{lat} development (e.g. caused by mode hops) are not described. Especially in the low power regime deviations occur, as will be shown in section 3.2.1.

Looking back to figure 3.3, the ground level is $BPP_0 \approx 1.8 \, \mathrm{mm \, mrad}$ and the thermal slope is $S_{th} \approx 0.1 \, \mathrm{mm \, mrad/K}$. This implies that for $\Delta T_{AZ} = 12.5 \, \mathrm{K}$, the thermally induced BPP_{lat} already reaches 50% of the total BPP_{lat}, confirming the strong impact of thermal lensing on the slow axis beam quality.

For every design-configuration of a BAL, the beam parameter product is measured on at least two devices per species (preferably more, if possible) for the same operating points, using the moving slit measurement station described in section 2.3. Afterwards, a linear regression is used to determine S_{th}, BPP_0 and the corresponding error ranges.

In the following sections, the linear model in equation 3.4 is used to quantify the impact of the influences under study, namely the lateral thermal lens *shape*, lateral index-guiding via dry etched trenches, strain fields, the lateral carrier distribution and the epitaxial design. By assessing how the ground level and BPP-slope are affected by these factors, the understanding of the broad area laser in-plane beam quality is improved and measures for increased radiance emerge.

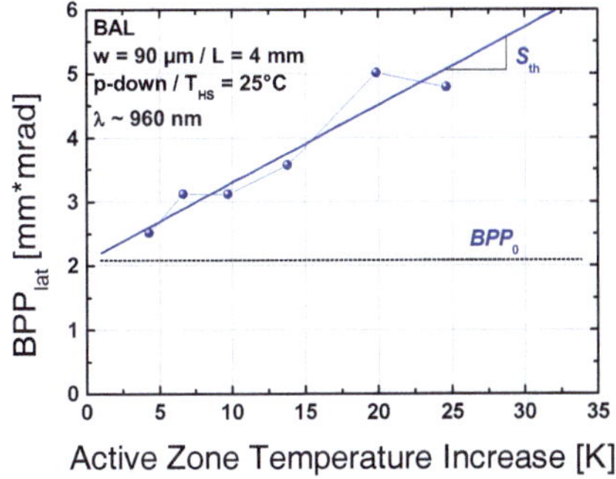

Figure 3.4: Linear empirical model of BPP_{lat} growth with internal temperature increase ΔT_{AZ}. The beam parameter product increases linearly with a thermal slope S_{th} and has a fundamental (non-thermal) ground level BPP_0.

3.2 Vertical design

From the listed factors in figure 3.1, the most basic thing to assess is the epitaxial design. It determines the vertical mode structure, the heat extraction and the efficiency of a BAL device (cf. section 2.1.3).

The most critical aspect with respect to this study is an optimized efficiency. The more efficient the design, the less thermal power needs to be removed and hence ΔT_{AZ} is reduced for equivalent power levels. However, in this section the focus is set on measures in the vertical design that *directly* improve the lateral beam quality. Specifically, the focus will be set on the vertical asymmetry, the chip geometry, the quantum well position and the modal gain.

3.2.1 Vertical asymmetry and thermal lens bowing

The main question here is: Does the asymmetry in the vertical structure influence the *lateral* modes? In [71], the use of an asymmetric design influenced the vertical-lateral near field aspect ratio in RWLs and allowed lateral single mode operation for broader ridge widths. However, in BALs multiple lateral modes are guided. But does the asymmetry influence their optical confinement? Another possible influence is the change in heat extraction in the sense that the degree of asymmetry may impact the thermal lens shape and hence influence lateral waveguiding.

In order to asses whether the vertical asymmetry has an impact on the BPP_{lat}, two groups of BALs with identical chip geometry were tested (simple gain guided BALs, cf. figure 3.5). The two designs 'ASLOC' and 'EDASLOC' were introduced in section 2.1.3. As noted in table 2.1 and visualized in figure 2.3, both designs differ in their asymmetry. In the 'ASLOC' design the fundamental mode expands symmetrically into the n- and p-waveguides. In contrast to that, the 'EDASLOC' design has minimal overlap in the p-part of the device. A thinned p-waveguide and an expanded GRIN n-waveguide cause the fundamental mode to expand widely into the n-region instead.

All lasers have cavity length of $L = 4\,\mathrm{mm}$, a stripe width of $w = 90\,\mu\mathrm{m}$ and a chip width of $w_{chip} = 1\,\mathrm{mm}$. Furthermore, the total thickness of the p-side (distance from quantum well to GaAs surface) equals $d_{p-side} = 1.4\,\mu\mathrm{m}$ for both designs.

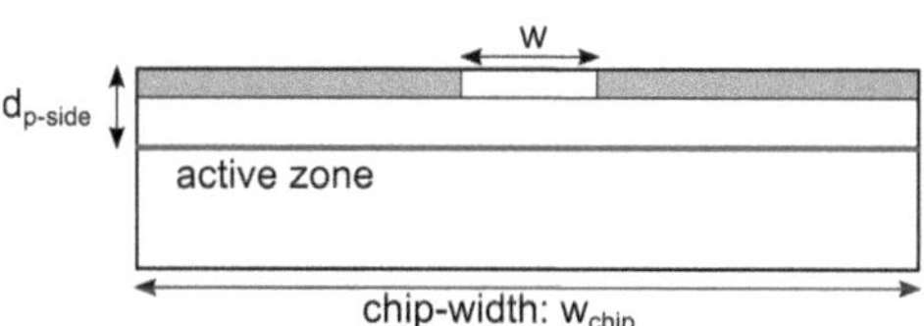

Figure 3.5: Sketch of the lateral-vertical laser geometry (x-y-plane), denoting the stripe width w, the chip-width w_{chip} and the total p-side thickness d_{p-side}. The stripe width w is defined by He$^+$ implantation of the cap layer (grey area).

The initial characterization of the BAL emitters encompasses LIV-measurements and spectral mappings. In figure 3.6(a), output power and efficiency are plotted as a function of current and a reduced threshold and increased efficiency for the ASLOC emitters is revealed. Afterwards, the lateral near and far field profiles (cf. figure 3.6(b)) are measured at selected output powers, using a moving slit configuration as described in section 2.3 and in [73]. The near fields show 'ear-like' side lobes, which are typical for gain-guided broad area emitters. At $P_{opt} = 10\,\mathrm{W}$, the near field waists are $w_{95\%} = 95\,\mathrm{\mu m}$ for ASLOC and $w_{95\%} = 86\,\mathrm{\mu m}$ for EDAS, respectively. The corresponding far field divergence is $\Theta_{95\%} = 10.8°$ for ASLOC and $\Theta_{95\%} = 12.8°$ for EDAS.

For every vertical design, the lateral near and far fields of three emitters were measured at the same operating points. The data was then averaged and the near field waists and far field angles were converted into BPP_{lat} values. As shown in figure 3.7, the BPP_{lat} increases with increasing output power and the data is fitted with a linear regression. The slopes read $\Delta BPP_{lat}/\Delta P = (0.20 \pm 0.01)\,\mathrm{mm\,mrad\,W^{-1}}$ for ASLOC and $\Delta BPP_{lat}/\Delta P = (0.34 \pm 0.03)\,\mathrm{mm\,mrad\,W^{-1}}$ for EDAS, indicating stronger beam quality deterioration for the design with the stronger asymmetry.

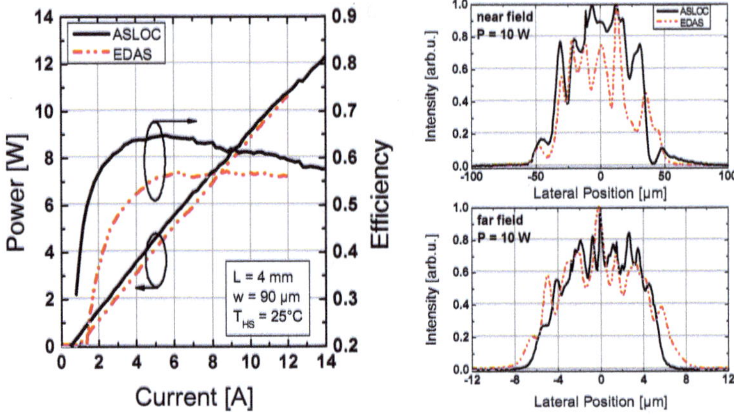

(a) Output power and conversion efficiency as func- (b) Near and Far field intensity profiles at
tion of current. $P_{opt} = 10\,\text{W}$.

Figure 3.6: Basic characterization of ASLOC (solid, black) and EDAS (dash-dotted, red) BALs. (a) cw LIV-measurement. (b) Representative near and far field profiles.

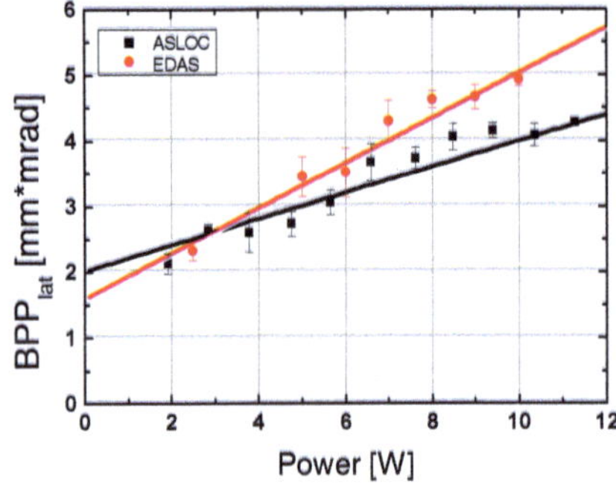

Figure 3.7: Lateral beam parameter product as function of output power of BAL emitters with ASLOC (black) and EDAS (red) vertical design. The error bars indicate the standard error of the mean and the solid lines represent a linear regression.

Vertical asymmetry and lateral mode confinement

How can the difference in BPP_{lat} presented in figure 3.7 be explained? In order to find a correlation between the vertical asymmetry and the slow axis beam quality, a simulation-based study was performed. It seeks to find a *direct* relation between the asymmetry and the confinement of the individual lateral modes.

Therefore a two-dimensional waveguide simulation was done for both structures, using the complex FEM solver from the simulation software FIMMWAVE. For each structure, the simulation domain contained the whole layer stack in the vertical (excluding carrier material) and 110 µm in the lateral direction. A 90 µm broad, rectangular shaped gain region inside the active area was used to emulate the injection stripe.

For both structures, the same lateral thermal lens was assumed (cf. figure 3.8(a)). It is the result of a FEM thermal simulation of the EDAS design in the case of p-side down mounting (operating point: $\Delta T_{AZ} = 18\,\text{K}$, $P_{opt} = 10\,\text{W}$, $I = 11.4\,\text{A}$, $U = 1.578\,\text{V}$, $I_{th} = 968\,\text{mA}$ and $R_s = 16.5\,\text{m}\Omega$). The FEM simulation (including the CuW carrier) was performed with JCMsuite and a detailed description of the heat source distribution and the boundary conditions is given in [73]. By using the $Al_xGa_{1-x}As$ refractive index function $n(x, \lambda, T)$ published by Gehrsitz $et\ al.$ [47], the temperature-profile was converted into a refractive index profile and implemented in the FIMMWAVE simulation.

The refractive index profile of the thermal lens is shown in figure 3.8(a). For the sake of simplicity, it was assumed that the thermal lens profile is constant in the vertical direction. The calculation gives the 2D waveguide modes up to mode number 12. The 9th mode for the EDAS design is shown in figure 3.8(b) as an example.

For both structures, all modes were evaluated in terms of confinement and losses. The result is summarized in figure 3.9. The vertical-lateral confinement factor $\Gamma_{x,y}$ is defined as follows:

$$\Gamma_{x,y} = \frac{\int^R P(s)ds}{\int P(s)ds}. \tag{3.5}$$

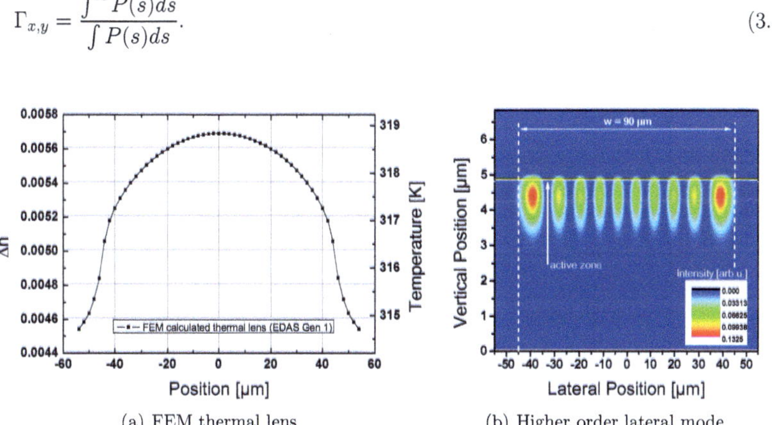

(a) FEM thermal lens   (b) Higher order lateral mode

Figure 3.8: Thermal lens input and a lateral mode calculated in FIMMWAVE simulation. (a) FEM-simulated thermal lens in 2 µm resolution (b) Example lateral mode (9th) for EDAS design.

Here, the integral in the numerator sums up the mode power P in the region of interest R (i.e. the quantum well underneath the 90 µm injection stripe). The mode losses are derived from the imaginary part of the mode effective index n_{eff}:

$$\alpha_{mode} = \frac{4\pi\mathrm{Im}(n_{eff})}{\lambda_0}. \tag{3.6}$$

In both designs, the low order modes have equal confinement. In absolute values, the ASLOC-modes are more strongly confined, but for both structures it holds that the higher the mode number, the greater the lateral mode expansion that is observed. Beginning with mode number 8, the confinement starts to decrease due to further mode expansion. This leads to reduced gain for the higher order modes, as depicted in figure 3.9(b). In figure 3.9(c) the difference in relative mode gain (normalized to fundamental mode) between ASLOC and EDAS is small for low order modes. However, beginning with mode number 9, the gain for the EDAS modes decreases faster than the ASLOC mode gain. For the highest order mode in this study (# 12), the absolute gain difference reaches 5.1%, with the EDAS mode showing a 22% reduction in relative gain compared to ASLOC. This leads to the conclusion that there is an intrinsic difference in the lateral mode selection that is caused by the asymmetry. Given the same thermal waveguide, the higher order modes in the EDAS design are suppressed more effectively. This could explain the low ground level BPP_0 that is observed in EDAS BALs (cf. figure 3.7 and 3.14(a)).

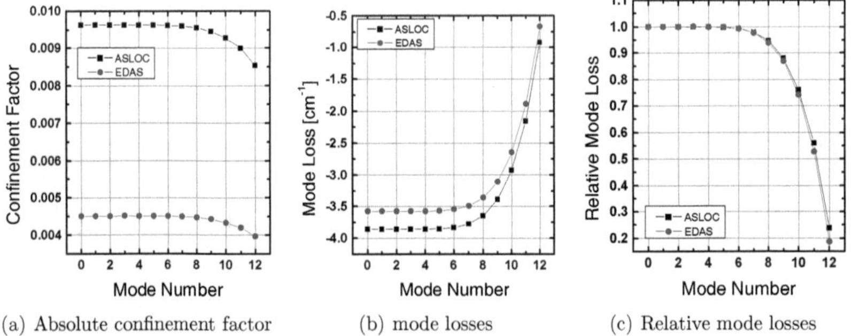

(a) Absolute confinement factor   (b) mode losses   (c) Relative mode losses

Figure 3.9: Evaluation of calculated lateral modes in ASLOC (black) and EDAS (red) structures. (a) ASLOC-modes have higher absolute confinement compared to EDAS-modes (b) The mode losses show constant gain for low order modes and a gain decrease for higher order modes (c) Almost no difference is seen in relative mode losses (normalized to fundamental mode losses).

However, even if the ground level is decreased, a faster BPP_{lat} degradation with ΔT_{AZ} is observed in EDAS BALs. Here a static analysis with fixed thermal lens is not appropriate. Moreover, due to the difference in the epitaxial structure, a different internal temperature distribution (thermal lens shape) and evolution is expected. Hence, a more direct way to assess these temperature effects is sought. In this study, a micro-thermography setup was used (cf. section 2.3.4) to gain information on the thermal lens shape inside broad area lasers. The thermal profiles for EDAS and ASLOC and their impact on BPP_{lat} will be analyzed in the following.

Thermal lens bowing in asymmetric structures

The diagram in figure 3.7 shows a higher BPP-slope of the EDAS-emitters with increasing power. Since the EDAS design is less efficient, its BALs have a higher junction temperature at any given power. For a fair comparison of both designs a plot of BPP_{lat} at equal temperature increase is necessary, according to the model in equation 3.4. In addition to the junction temperature, the shape of the lateral-vertical temperature distribution is of great interest, since it determines the thermal waveguide. Hence, a micro-thermography technique was used to determine both the active zone temperature increase and the shape of the thermal lens. A representative 2D temperature map for an ASLOC BAL at $P_{opt} = 10\,\mathrm{W}$ is shown in figure 3.10(a).

For the analysis of the BAL thermal lens, the pixel row that includes the highest temperature in the center of the chip is extracted from the temperature map. Due to the poor spatial resolution, this pixel row contains temperature information that is an average of all epitaxial layers, including the active area. A series of these lateral temperature profiles for multiple output powers is shown in figure 3.10(b). With increasing output power the overall chip temperature, measured at the edge regions with $|x| \geq 400\,\mathrm{\mu m}$, increases up to 39 °C at $P_{opt} = 13\,\mathrm{W}$. Furthermore, the thermal lens in the chip center increases in height and steepness. In order to quantify these observations, the thermal profiles from 2 emitters per species are evaluated with respect to the relative temperature increase ΔT_{AZ}, the center-to-edge temperature difference ΔT_{ce} (for definition see figure 3.10(b)) and the thermal lens bowing.

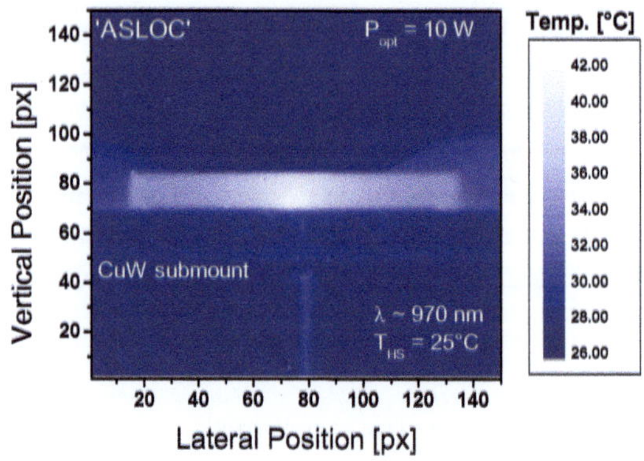

(a) Temperature map derived from thermal camera images of exemplary ASLOC emitter at $P_{opt} = 10\,\text{W}$.

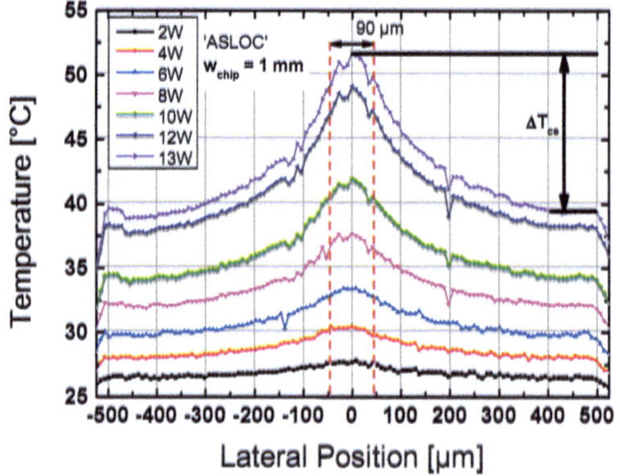

(b) Lateral Temperature profiles (along active region, extracted from (a)) for output powers up to $P_{opt} = 13\,\text{W}$.

Figure 3.10: Exemplary data from micro-thermography measurements. (a) Temperature map shows ASLOC laser chip on submount during cw-operation. The calibration shown gives correct temperatures solely for the GaAs material. (b) Lateral temperature profiles extracted from the pixel-row in (a) that contains the maximum temperature (i.e. includes the active zone). The vertical dashed lines indicate the edges of the current aperture.

Before this analysis is presented, a few remarks on the validity of these profiles should be made. First, the thermal lens profiles in figure 3.10(b) are much broader than the simulated profiles [73] and those from earlier photo-luminescence micro-probe experiments [74]. The afore mentioned broadening effect of infrared radiation is the main cause for this observation. In addition, steep lateral gradients might not be resolvable due to the limited camera resolution. Secondly, the camera detector chip has a few pixel failures (e.g. the pixel at $200\,\mu\text{m}$ in figure 3.10(b) for $P_{opt} \geq 8\,\text{W}$), that add an artificial, systematic error to any fitting procedure applied to the temperature data. However, the micro-thermography technique still allows an analysis of relative differences between thermal profiles taken from different vertical structures.

To describe the shape of the thermal lens, a bowing parameter B_2 is introduced that is taken from a quadratic fit of the central temperature profile in the range $|x| \leq 60\,\mu m$, as depicted in figure 3.11.

The evaluation of the camera-measured temperature increase is shown in figure 3.12(a). Compared to the ΔT_{AZ} that was determined via the wavelength-shift, the camera signal gives a slightly lower temperature, which can be explained by the broadening effect that redistributes the IR-photons from the central region to the neighboring edge-regions of the chip. The shape of the temperature curve, however, is in good agreement with the curve that was generated via the standard method (wavelength shift), following a quadratic increase with nearly the same coefficients except for the y-axis intercept.

In figure 3.12(b), the lateral beam parameter product of the EDAS emitters is depicted as function of ΔT_{AZ}. The graph compares two possibilities of data evaluation, which will be discussed here briefly. The open symbols with error bars (standard error of the mean) are based on the BPP_{lat} measurement of 3 EDAS emitters and for every operating point the temperature increase was derived according to equation 3.3 (wavelength shift). A linear fit of this data gives a thermal slope $S_{th} = \Delta BPP_{lat}/\Delta T_{AZ} = (0.21 \pm 0.02)\,\mathrm{mm\,mrad\,K^{-1}}$ and a ground level $BPP_0 = (1.3 \pm 0.3)\,\mathrm{mm\,mrad}$. The full symbols represent a function that combines two fitting curves. For any given output power P_{opt}, the BPP_{lat} can be derived using the linear regression shown in figure 3.7 and the corresponding ΔT_{AZ} is derived using the (camera-measured) quadratic fit-curve given in figure 3.12(a).

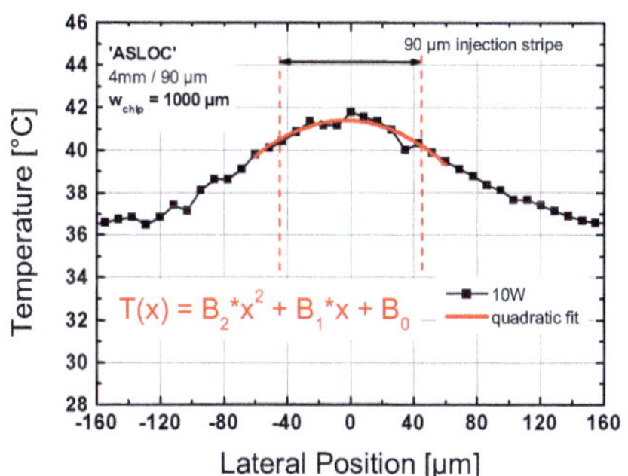

Figure 3.11: Lateral temperature profile for ASLOC BAL at $P_{opt} = 10\,\mathrm{W}$ in the range of $|x| \leq 160\,\mu m$. A quadratic fit for $|x| \leq 60\,\mu m$ gives the quadratic coefficient B_2, which is used to quantify the thermal lens bowing.

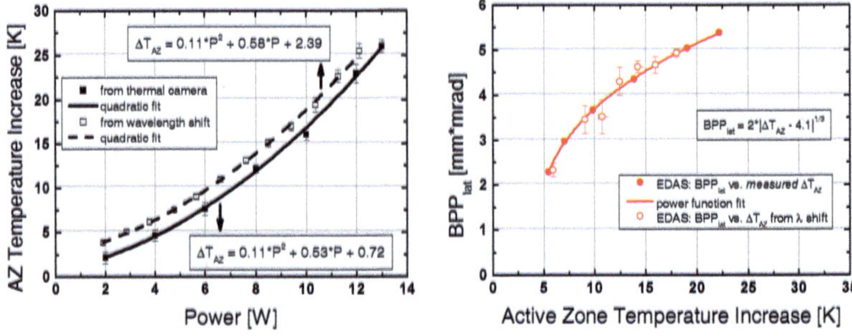

(a) Measured ΔT_{AZ} as function of output power for exemplary ASLOC emitter.

(b) Lateral beam parameter product as function of ΔT_{AZ}.

Figure 3.12: Evaluation of the thermal camera data and comparison to the results of the wavelength shift method. (a) The comparison of camera-measured and wavelength-drift determined ΔT_{AZ} shows good agreement. Both methods yield a quadratic temperature increase, with the latter having $\approx +1.7\,\text{K}$ offset. (b) BPP_{lat} as function of camera-measured ΔT_{AZ} follows a power function, that also fits the original $BPP_{lat}(\Delta T_{AZ})$ data within measurement uncertainty.

On the one hand, this method decouples the beam quality measurement from the thermal camera measurement and hence enables data interpolation, even if BPP_{lat} and ΔT_{AZ} have not been measured at the same power levels. On the other hand, it relies on fitting curves, that can deviate from the measured data. The resulting BPP graph is well described by a power function of the form $BPP_{lat} = A \times |\Delta T_{AZ} - x_c|^p$, with $A = (2.00 \pm 0.05)\,\text{mm}\,\text{mrad}\,\text{K}^{-p}$, $x_c = (4.1 \pm 0.1)\,\text{K}$ and $p = 0.33 \pm 0.01$. The comparison in figure 3.12(b) reveals that the actual BPP_{lat} function is more complex than the simple linear model from equation 3.4 implies. In the regime $\Delta T_{AZ} \leq 10\,\text{K}$, the power function is more accurate, since it reproduces the drop in BPP_{lat} for low injection levels. Here, only lower order modes are excited and the BPP approaches its fundamental limit $BPP_{lim} = \lambda/\pi$. For high injection levels and $\Delta T_{AZ} > 10\,\text{K}$, the power function is well approximated with a linear increase. Hence, a linear model, as described in section 3.1, is still reasonable - especially if the focus is set on the high-power regime, which is the case in this study. However, the linear fits must be interpreted with care.

Finally, the two vertical designs in this test are compared by means of temperature increase and thermal lens bowing. The temperature elevation between the center and edge of the chip is comparable for both structures (cf. figure 3.13), with EDAS having a slightly higher T_{ce} at increased ΔT_{AZ}.

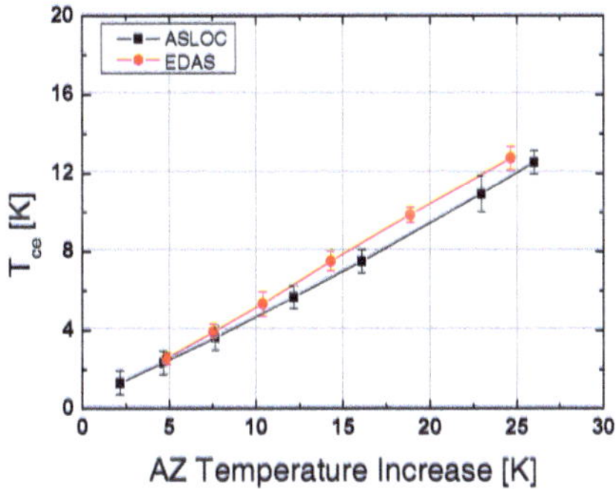

Figure 3.13: Center to edge temperature increase T_{ce} as function of ΔT_{AZ} for the ASLOC and EDAS designs.

The increased slope of the BPP_{lat} as function of power in EDAS BALs is also reflected in the $BPP_{lat}(\Delta T_{AZ})$ diagram, as can be seen in figure 3.14(a) and table 3.1. Compared to ASLOC BALs, the thermal slope S_{th} is doubled in EDAS BALs. In figure 3.14(b), the thermal lens bowing for both designs is depicted. Here, the EDAS BALs show an increased bowing value B_2 and an accelerated bowing increase with ΔT_{AZ}, since $\Delta B_2/\Delta T$ is 40% higher than for to ASLOC BALs.

This observation is consistent with phenomenological descriptions of thermal blooming [46]. An increasingly bowed thermal lens gives a stronger waveguide that can guide more lateral modes and causes the BPP_{lat} to deteriorate.

The correlation in figure 3.14 shows that the vertical design of a BAL influences the thermal distribution inside the laser chip, which causes a change of the thermal slope S_{th}. In this study, all BALs are soldered p-side down, so that the generated heat flows through the p-side of the vertical structure. Since the total p-side thickness d_{p-side} was not changed, a closer look at the individual layers seems worthwhile.

parameter	unit	ASLOC	EDAS
S_{th}	[mm mrad K^{-1}]	0.09 ± 0.01	0.18 ± 0.02
BPP_0	[mm mrad]	2.4 ± 0.1	1.6 ± 0.3
$\Delta B_2/\Delta T$	[°C/µm^{2}K]	$-(3.3 \pm 0.1) \times 10^{-5}$	$-(4.6 \pm 0.2) \times 10^{-5}$

Table 3.1: Summarized parameters for linear BPP_{lat} fit and thermal lens bowing.

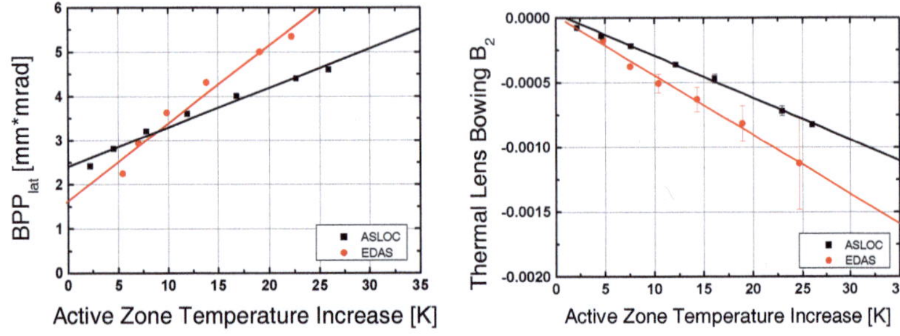

(a) Comparison of ASLOC and EDAS design in BPP_{lat} vs. ΔT_{AZ} plot.

(b) Thermal lens bowing as function of ΔT_{AZ}.

Figure 3.14: Slow axis BPP (a) and thermal lens bowing (b) as functions of ΔT_{AZ} for ASLOC and EDAS epitaxial design. For EDAS devices the bowing increase is stronger, which correlates to a larger thermal slope S_{th}.

In figure 3.15(a), the thermal conductivity of $Al_xGa_{1-x}As$ is plotted as function of the aluminum content [75, 76]. In addition, the aluminum contents of the p-waveguide and p-cladding for ASLOC and EDAS are noted, showing that 77% of the EDAS GRIN close to the active zone consists of material that has a lower thermal conductivity than the ASLOC p-waveguide. In figure 3.15(b) the formula for $\lambda_{th}(x)$ given in [75] is used to derive a vertical profile of the p-side thermal conductivity.

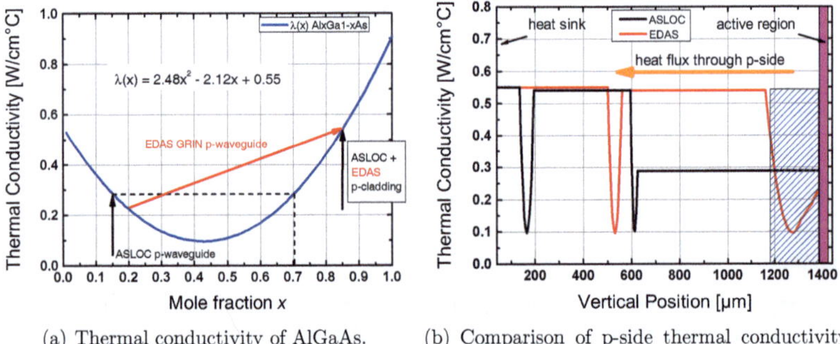

(a) Thermal conductivity of AlGaAs.

(b) Comparison of p-side thermal conductivity vertical profiles.

Figure 3.15: Illustration of differences in thermal properties of ASLOC and EDAS p-doped epitaxial layers. (a) Thermal conductivity λ_{th} of $Al_xGa_{1-x}As$ as function of aluminum mole fraction, taken from [75]. The aluminum content of p-waveguide and -cladding are noted for both designs. (b) Depth profile of λ_{th} from cap to active area. Right above the quantum well (blue shaded area), the EDAS design has significantly reduced thermal conductance.

In the 200 nm thick area right above the quantum well (blue shaded zone in figure 3.15(b)), the conductivity is significantly reduced for EDAS emitters. In this zone, the thermal resistance of ASLOC emitters amounts to only 53.7% of the EDAS value. For an extended 500 nm zone, the two designs differ less strongly. Here, the ASLOC thermal resistance is 93% of the EDAS value. However, the calculated thermal resistance of the complete p-layer stack reads $R_{th,ASLOC} = 0.113\,\mathrm{K\,W^{-1}}$ and $R_{th,EDAS} = 0.103\,\mathrm{K\,W^{-1}}$, showing that EDAS has a lower total p-side thermal resistance. In both cases, the calculated p-layer R_{th} is only $\approx 5\%$ of the total (measured) thermal resistance, but since the soldering process is identical for both designs, the main difference is expected to be chip-internal.

Now, for the increased thermal lens bowing in the EDAS BALs, a possible explanation emerges. The 200 nm thick region above the quantum well could play a large role in the formation of a strong thermal lens due to its low conductance and its proximity to the active region, where most of the heat is produced.

Conclusions

To summarize this section, it is noted that for BAL emitters with the same chip geometry, the vertical structure influences the lateral-vertical temperature distribution. A key figure of merit here is the thermal lens bowing, which is directly correlated to the BPP_{lat} increase with temperature. The design of the p-side layer stack, specifically the aluminum content and the use of graded layers, was found to have a strong influence on the thermal lens shape. It was shown that an extreme asymmetric design (EDAS) has increased the thermal lens bowing and therefore had poorer slow axis beam quality than the more conventional ASLOC design with symmetric waveguide layers.

An analysis of the individual lateral modes showed a slightly greater mode discrimination for higher order modes in the more asymmetric EDAS design. This could explain the reduced BPP_0 in that structure. In general, however, the factors influencing BPP_0 remain poorly understood.

3.2.2 Importance of quantum well position and chip geometry

In this section, the focus is set on the chip geometry itself. Two questions wait to be answered here. First: Is it beneficial to increase the distance between the quantum well and the chip surface d_{QW} in order to allow 'chip-internal' heat spreading for homogenization of the lateral thermal profile? And second: What happens to the BPP_{lat} if the laser chip is more compact, i.e. the chip-width w_{chip} is reduced?

In general, there are two considerations regarding d_{QW} in p-side down mounted BALs. On the one hand, d_{QW} should be as small as possible for improved thermal resistance. In addition, p-GaAs has poor electric conductance compared to metals or n-GaAs while the series resistance is aimed to be as low as possible. On the other hand, however, the quantum well needs to be protected from potential strain fields or non-uniformities caused by the soldering process (even if planar structures are used) and the p-cladding needs to be thick enough to avoid leakage of the fundamental mode into the highly doped p-cap layer, which otherwise would increase the internal losses dramatically.

The findings in section 3.2.1 imply that the impact of the p-side on the thermal distribution is another important consideration, because the slow axis beam quality is directly correlated to it. In order to find an answer to the first question, the comparison of two extreme cases is done by testing p-side up and p-side down mounted BALs of the same type. In the case of p-side down mounting, $d_{QW} = d_{p-side} = 1.4\,\mu\mathrm{m}$, while for p-side up mounting $d_{QW} = d_{n-side} + d_{substrate} \approx 128\,\mu\mathrm{m}$, yielding two orders of magnitude difference.

In figure 3.16, the two configurations are compared in a thermal simulation and in the BPP_{lat} vs. ΔT_{AZ} diagram. The simulation in figure 3.16(a) was done with the software JCM Suite and the description of the model and input parameters can be found in [73]. Here, the lateral thermal profiles for p-side down and p-side up mounted devices are compared at equal temperature increase $\Delta T_{AZ} = 18\,\mathrm{K}$. Since the thermal resistance for p-up devices, $R_{th,up} = 5\,\mathrm{K\,W^{-1}}$, is more than twice that of p-down devices with $R_{th,down} = 2.4\,\mathrm{K\,W^{-1}}$, the temperature increase is reached at a lower output power in the p-up devices.

However, as can be seen in the edge regions of the thermal profile, the p-down BALs show a steep temperature gradient around $x = \pm\,50\,\mu\mathrm{m}$ compared to their p-up siblings, which show a slower varying lateral temperature gradient. Since the near field expansion for purely gain guided BALs (indicated by the dotted vertical line) is larger than $\pm\,50\,\mu\mathrm{m}$, an impact on the slow axis BPP is expected.

In figure 3.16(b), the BPP_{lat} is depicted as function of internal temperature increase. The thermal slope for p-side down mounted BALs is $S_{th,down} = (0.21 \pm 0.02)\,\mathrm{mm\,mrad/K}$ and for p-side up mounted BALs the slope is less than half, with $S_{th,up} = (0.09 \pm 0.02)\,\mathrm{mm\,mrad/K}$. Both configurations share the same $BPP_0 = 1.3\,\mathrm{mm\,mrad}$, which is reasonable because the BPP ground level is considered to be the non-thermal part of the diode laser's beam quality. This result implies that the mounting does not intrinsically add any parasitic effects that deteriorates BPP_0. The difference in the thermal slope is attributed to the different shapes of the thermal lens profiles in the stripe edge, with the p-side up profile without the abrupt step at $\pm\,50\,\mu\mathrm{m}$ being advantageous.

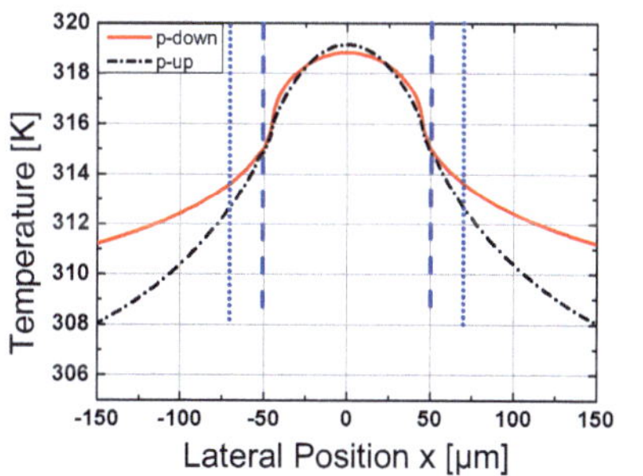

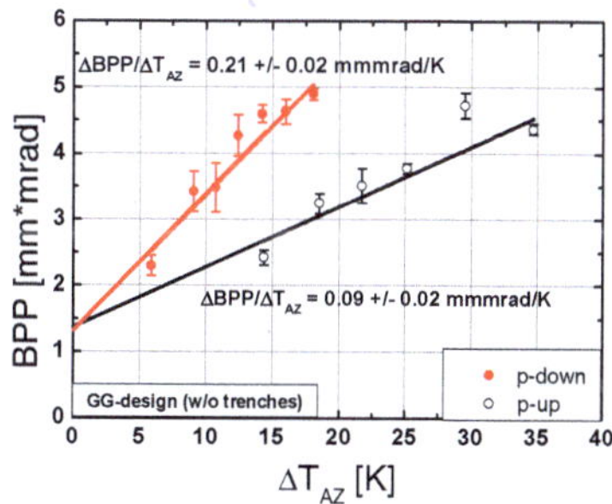

(a) JCM Suite simulation of lateral thermal lens profile for p-up and p-down EDAS BAL.

(b) Comparison of p-side up and p-side down mounted BALs in BPP_{lat} vs. ΔT_{AZ} diagram.

Figure 3.16: Test of the impact of the quantum well position with respect to the heat sink. (a) FEM-simulation of lateral thermal profile along active zone in p-side down (red, solid curve) and p-side up (black, dash-dotted curve) mounted EDAS broad area lasers. The vertical lines indicate the near field expansion for devices with (dashed) and without (dotted) index guiding trenches. (b) Lateral BPP as function of temperature increase shows reduced thermal slope for p-side up mounted emitters.

So the first question is answered positive, at least in this test where d_{QW} was increased by a factor of $\approx 100x$. However, since p-side up mounting is no option for high power laser operation due to poor heat extraction, other possibilities for thermal homogenization in p-side down mounted BALs are sought.

During this study, the EDAS design was adapted in order to test whether increased p-side thickness would help to reduce the thermal lens bowing and lateral BPP. The changes in the vertical structure in EDAS generation two compared to the first generation are summarized in table 3.2. The new design is available with a SQW and a DQW, so that the effect of a modal gain variation can be assessed (see section 3.2.3). However, the major difference in EDAS generation two is found in the p-side epitaxial layers. First, the sub-contact layer (highly p-doped layer beneath the p-cap) thickness was increased by 100%, from 400 nm to 800 nm. Second, the p-cladding thickness was increased by 33%, from 600 nm to 800 nm. In addition, 2% phosphorous was added to the $Al_{0.85}Ga_{0.15}As$ matrix in the p-cladding for strain compensation. Apart from these changes, a few smaller adaptions in the waveguide aluminum content and doping profile were made. However, the focus is set on the impact of the increased p-side thickness d_{p-side}, which rose from 1.4 μm to 2 μm (+42%).

Another important change that should be noted here is the reduction in chip width w_{chip} in the BAL process for EDAS generation two. It was reduced from 1 mm to 0.5 mm, such that the comparison encompasses two simultaneous geometrical changes.

layer	EDAS Gen 1	EDAS Gen 2 (DQW)	EDAS Gen 2 (SQW)
sub-contact	$0.4\,\mu$m GaAs	$0.8\,\mu$m GaAs	
p-cladding	$0.6\,\mu$m $Al_{0.85}Ga_{0.15}As$	$0.8\,\mu$m $Al_{0.85}Ga_{0.15}As_{0.98}P$	
p-wg. GRIN	$Al_xGa_{1-x}As$: $0.85 \geq x \geq 0.2$	$Al_xGa_{1-x}As$: $0.85 \geq x \geq 0.15$	
active zone	InGaAs DQW and GaAsP barriers		InGaAs SQW GaAs spacer
n-wg. GRIN	$2.6\,\mu$m $Al_xGa_{1-x}As$: $0.2 \leq x \leq 0.35$	$2.6\,\mu$m $Al_xGa_{1-x}As$: $0.15 \leq x \leq 0.3$	$3.4\,\mu$m $Al_xGa_{1-x}As$: $0.15 \leq x \leq 0.3$
n-cladding	$1.5\,\mu$m $Al_{0.35}Ga_{0.65}As$	$1.5\,\mu$m $Al_{0.3}Ga_{0.7}As$	$1\,\mu$m $Al_{0.3}Ga_{0.7}As$

Table 3.2: Details of the epitaxial structure for the EDAS generations in this study. The emission wavelength has been adapted to be $\lambda = 960\,$nm in EDAS Gen 2.

In figure 3.17, the performance of BALs from the two consecutive EDAS generations is presented. In figure 3.17(a), the power and efficiency curves are shown as function of current. The maximum efficiency η_{max} could be increased by 5% in generation two.

In figure 3.17(b), the impact on the BPP_{lat} is shown. The increased p-side thickness d_{p-side} and the reduced chip width w_{chip} result in an improved thermal slope S_{th}, which decreases by 33% from $(0.21 \pm 0.02)\,$mm mrad/K to $(0.14 \pm 0.01)\,$mm mrad/K. In addition, a slight reduction (-15%) in the BPP ground level is observed from $(1.3 \pm 0.3)\,$mm mrad to $(1.1 \pm 0.1)\,$mm mrad.

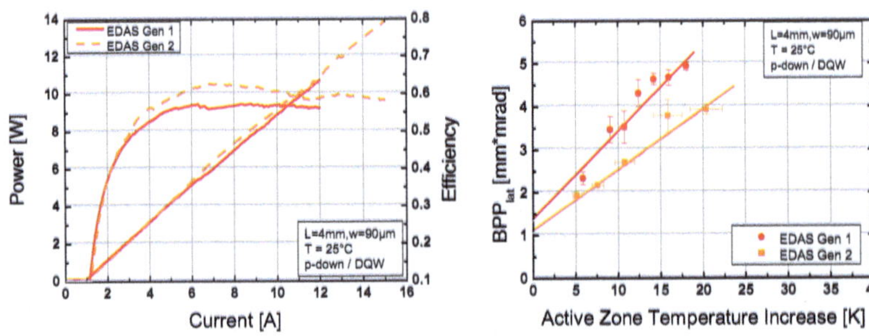

(a) Power and efficiency as function of current.

(b) BPP_{lat} as function of ΔT_{AZ} (determined from wavelength drift).

Figure 3.17: Comparison of power, efficiency and BPP_{lat} for DQW BAL emitters from EDAS generations one and two.

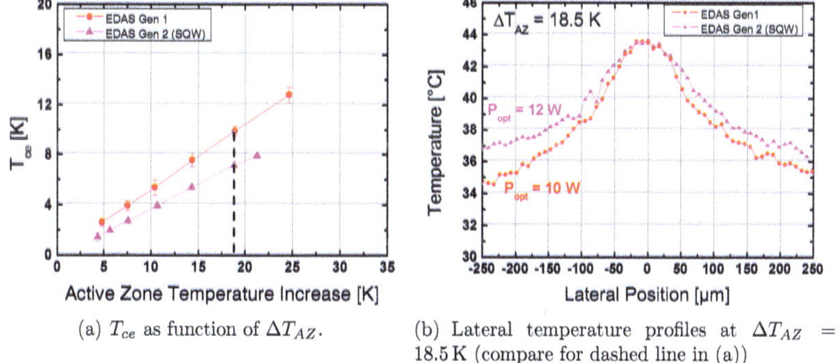

(a) T_{ce} as function of ΔT_{AZ}.

(b) Lateral temperature profiles at $\Delta T_{AZ} = 18.5\,\text{K}$ (compare for dashed line in (a))

Figure 3.18: Center to edge temperature increase for EDAS generations in test and comparison of lateral temperature profiles at fixed ΔT_{AZ}. In EDAS Gen 2, the thermal lens shows a slightly reduced bowing.

In order to analyze the changes in the thermal distribution, SQW EDAS emitters from generation two were assessed with thermal camera imaging. The results presented in figure 3.18 and figure 3.19 are based on a two emitter average. Regarding the center to edge temperature increase T_{ce} (cf. fig. 3.18(a)), it is not surprising that the BALs from generation two have constantly lower values, since the chip width was halved. However, the slope $\Delta T_{ce}/\Delta T_{AZ}$ is reduced as well, dropping by -25% from (0.52 ± 0.01) to (0.39 ± 0.01). In figure 3.18(b), the lateral temperature profiles for equal temperature increase are shown. Here, the bowing for EDAS generation 2 is slightly lower compared to generation one.

In figure 3.19(a), the BPP_{lat} is shown as function of ΔT_{AZ}. As in the comparison in figure 3.17(b), the BPP is reduced in EDAS generation two. Here, the decrease in BPP_0 is stronger compared to the DQW emitters: -39% down to $(0.8 \pm 0.1)\,\text{mm}\,\text{mrad}$. The thermal slope reduction is comparable to the previous observation and yields $S_{th} = (0.15 \pm 0.01)\,\text{mm}\,\text{mrad/K}$ for the SQW emitters of EDAS generation two. As shown in figure 3.19(b), the thermal lens bowing is only slightly improved and is partly within the error range of the bowing in EDAS generation one. The bowing increase is also slightly reduced from $-(4.6 \pm 0.2) \times 10^{-5}\,^{\circ}\text{C}/\mu\text{m}^2\text{K}$ to $-(4.1 \pm 0.1) \times 10^{-5}\,^{\circ}\text{C}/\mu\text{m}^2\text{K}$.

On this basis, the following conclusions are drawn. The 'compactification' of the laser chip (i.e. p-side thickness increase and chip width reduction) in EDAS generation two leads to an improved slow axis beam quality. Partly due to a reduction in thermal slope S_{th} and mostly due to a reduction in the BPP ground level BPP_0. The evaluation of thermal camera images showed a slight improvement of the thermal lens bowing, which is consistent with a small thermal slope decrease. In addition, the center to edge temperature elevation T_{ce} is reduced and grows more slowly in the compact chip design.

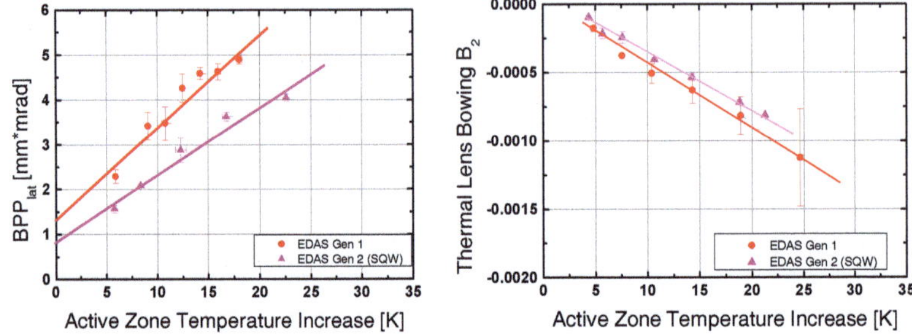

(a) BPP_{lat} as function of ΔT_{AZ} (determined from wavelength drift)   (b) Thermal lens bowing B_2 as function of ΔT_{AZ}

Figure 3.19: Comparison of EDAS generations after evaluation of thermal camera images. (a) Beam parameter product as function of temperature increase. (b) Thermal lens bowing.

Hence, the chip size reduction leads to a lateral homogenization of the temperature distribution (cf. fig 3.18(b)), which also smoothens the temperature profile in the critical area below the injection stripe, leading to slightly improved S_{th}. The reduction in the BPP_0 can be seen as a sign for improved waveguide quality, either via chip width reduction or as a consequence of increased p-side thickness.

3.2.3   Modal gain and filamentation

Among the list of factors for beam quality degradation (cf. fig. 3.1), filamentation has been studied intensively in the past [41, 70]. In this subsection, the importance of filamentation for BPP_{lat} and its suppression via a modal gain decrease will be discussed.

Filamentation is defined as the self-organized formation of local intensity peaks inside a broad gain medium and is often seen as a major threat to slow axis beam quality in broad area lasers and amplifiers. It is a long known phenomenon, but its impact on BPP_{lat} in modern high power BALs is unclear. Hence, an experimental investigation was performed here, based on a comparison of two vertical structures with different modal gain.

The basis for this study is set in a publication by G.C. Dente [77], where the filamentary behavior in broad area lasers is modeled as follows:

$$E(x, z) = E_0(x, z) + f \cdot \exp(\gamma z) \cdot \sin(k_f x). \tag{3.7}$$

In equation 3.7, the electric field $E(x, z)$ consists of a steady state field $E_0(x, z)$ and a sinusoidal field perturbation f with spatial frequency k_f. This perturbation grows as the field propagates along the resonator (z-)axis with a growth rate γ (filament gain). For this filament gain, a maximum value is given in [77]:

$$\gamma_{max} = \frac{\sqrt{\alpha_H + 1} - 1}{2} \cdot \frac{I_0/I_s}{1 + (I_0/I_s)} \cdot (\alpha_{int} + \alpha_m),$$ (3.8)

where α_H denotes the linewidth enhancement factor, α_m the mirror losses, α_{int} the internal losses and $I_0 = |E_0|^2$ the steady state intensity. The so-called saturation intensity I_s is defined as:

$$I_s = \frac{\eta_i V_{gap}}{\Gamma \left(\delta G/\delta j\right)},$$ (3.9)

with η_i denoting the internal efficiency, Γ the confinement factor, $\delta G/\delta j$ the differential gain and V_{gap} the band gap energy.

By minimizing the filament growth rate γ, broad area lasers are expected to operate with more homogeneous (i.e. less modulated) field profiles. In [40], it was shown that the beam quality (described by the beam propagation ratio $M_{lat}^2 = BPP_{lat} \cdot \pi/\lambda$) can be expressed as the square root of the sum of three components, each representing a laser beam attribute (cf. equation 2.12). These attributes are, to repeat, the transverse power density distribution, the beam wavefront and its transverse distribution of coherence. Hence, the stronger a beam is modulated, the higher the BPP_{lat} and therefore a reduction in γ is expected to be beneficial.

From equation 3.8, a few options for decreasing γ_{max} emerge: Reducing the linewidth enhancement α_H and the internal losses α_{int} would be helpful. However, in this investigation the saturation intensity I_s will be increased via a decrease in differential gain $\Delta G/\Delta j$. For this, a change in the number of quantum wells is sufficient and BAL emitters from EDAS generation two with single and double quantum wells (see table 3.2) are ideal candidates for comparison, since the vertical asymmetry and chip geometry are not changed.

In figure 3.20, data from basic laser characterization is shown. Here, a series of uncoated and unmounted BALs with $w = 100\,\mu\text{m}$ was tested in length-dependent and pulsed LIV-measurements to obtain the modal gain parameter Γg_0 and the transparency current density j_{trans}. A detailed description of the evaluation procedure is given in [34].

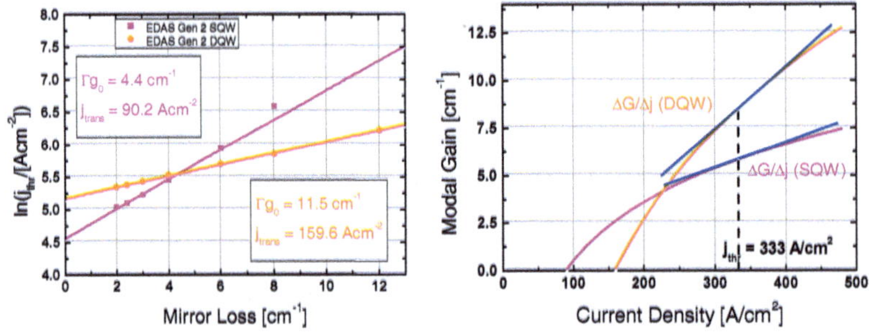

(a) Logarithmic threshold current density as function of mirror losses.

(b) Modal gain vs. current density.

Figure 3.20: Basic characterization of EDAS Gen 2 SQW (magenta) and DQW (orange) BAL emitters. (a) Modal gain parameter Γg_0 and transparency current density j_{trans} inferred from length-dependent measurements of $w = 100\,\mu m$ BALs in pulsed regime. (b) Modal gain curves as function of current density in logarithmic gain model, using parameters derived from (a). The differential gain $\Delta G/\Delta j$ at threshold is indicated as a blue line.

In figure 3.20(a), it is shown that the addition of a further quantum well to the SQW structure increases Γg_0 from $4.4\,cm^{-1}$ to $11.5\,cm^{-1}$ and j_{trans} increases from $90\,A\,cm^{-2}$ to $160\,A\,cm^{-2}$. These parameters are used to plot the gain-current function for both EDAS designs (logarithmic gain model [78]) and infer the differential gain $\Delta G/\Delta j$ at threshold, as presented in figure 3.20(b).

Now, according to equation 3.9, the saturation intensity is inversely proportional to the differential gain, which reads $\Delta G/\Delta j = 0.013\,cm\,A^{-1}$ and $0.033\,cm\,A^{-1}$ for the SQW and DQW designs, respectively. This yields a 2.5x increase in I_s for the SQW design, which results in a 2.2x reduction in maximum filament gain γ_{max}. In this calculation, the linewidth enhancement was assumed to be $\alpha_H = 2$ [79]. A summary of all parameters is given in table 3.3.

As shown in figure 3.21, the optical performance of both designs is comparable up to $P_{opt} = 12\,W$. The output is limited by rapid thermal rollover with $P_{max,SQW} = 14\,W$ and $P_{max,DQW} = 19\,W$. The laser facets are not damaged since the measured LIV curves are fully reproducible, so that the passivation technology guarantees a facet failure power $P_{COMD} > 19\,W$.

The facet coatings were adapted to yield the same threshold current density $j_{thr} \approx 333\,A\,cm^{-2}$ (as seen in fig. 3.21), with rear facet reflectivity $R_r > 95\,\%$ for both designs and front facet reflectivity $R_f = 2\,\%$ and $R_f = 0.8\,\%$ for SQW- and DQW design, respectively.

parameter	unit	DQW	SQW
α_H	-	2	2
Γg_0	[cm^{-1}]	11.5	4.4
α_{int}	[cm^{-1}]	0.69	0.38
α_m	[cm^{-1}]	6.09	4.08
η_i	-	0.98	1
$\Delta G/\Delta j$	[cm/A]	0.033	0.013
I_s	[W cm^{-1}]	39.4	99.8
γ_{max}	[rel.u.]	1	0.46
BPP_{lat} (10 W)	[mm mrad]	3.5	3.6
c_{NF} (10 W)	-	1.1	0.58

Table 3.3: Summarized material parameters and filamentation properties of EDAS Gen 2 SQW and DQW designs under study.

But even at equal threshold current density, the change in the number of quantum wells causes a difference in carrier density per well, which could have an impact on BPP_{lat} (see section 3.4). Unfortunately, it is not clear how pronounced this difference actually is. This depends on the coefficients for spontaneous and amplified emission inside the active material [19]. The main difference between both designs, however, is still in the modal gain, which will be used as a diagnostic tool in this experiment.

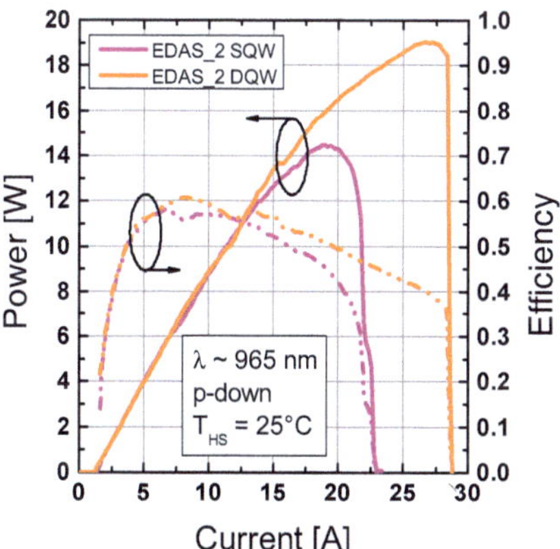

Figure 3.21: Optical output power P_{opt} and conversion efficiency η_c as function of current for p-side down mounted $w = 90\,\mu$m BALs with $L = 4\,$mm from EDAS Gen 2 SQW and DQW designs. The maximum output power P_{max} is limited by rapid thermal rollover.

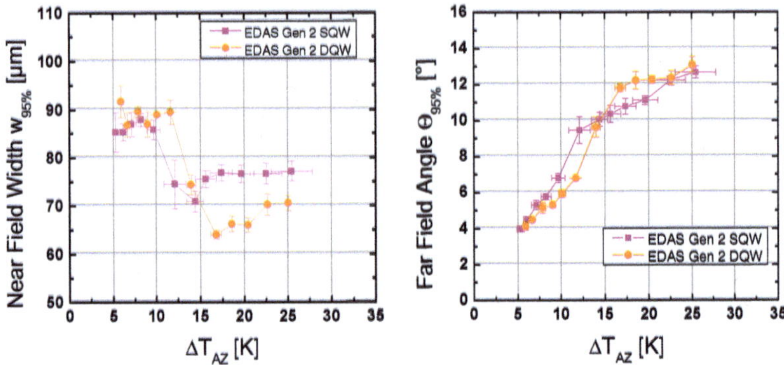

(a) Averaged near field waist vs. ΔT_{AZ} (b) Averaged far field divergence vs. ΔT_{AZ}

Figure 3.22: Near- (a) and far field (b) dimensions at 95 % power content as function of temperature increase for EDAS Gen 2 BALs with SQW and DQW gain regions.

In figure 3.22, the averaged near field widths and far field angles (4 emitters per species) are presented as function of ΔT_{AZ}. The DQW BALs have a slightly increased near field and a reduced far field divergence for $\Delta T_{AZ} < 15\,\mathrm{K}$. In both designs, the near field width shrinks with increasing temperature, but for $\Delta T_{AZ} > 15\,\mathrm{K}$ the DQW near field width is narrower than the SQW near field. Simultaneously, the DQW far field angle increases beyond that of the SQW design.

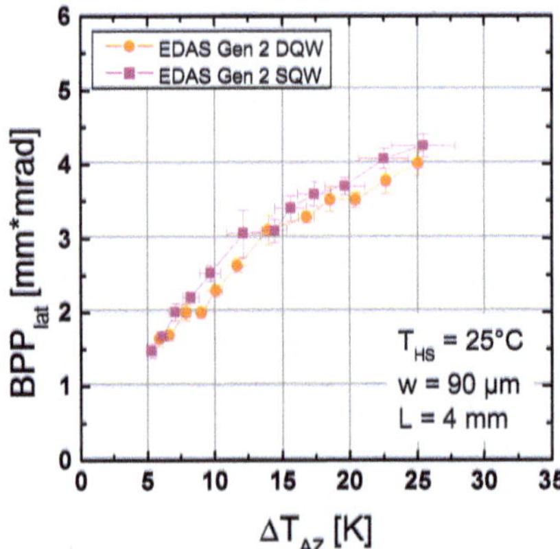

Figure 3.23: SQW vs. DQW: Compiled BPP_{lat} as function of ΔT_{AZ}. Although the filament gain γ_{max} is less than half, no clear improvement is seen for the SQW design.

After compiling the near and far field data into BPP_{lat} (see figure 3.23), no clear difference between the SQW and the DQW design is seen, although the filament gain is less than half for the SQW emitters. In fact, the beam quality of the DQW emitters is slightly better.

Hence, the conclusion from this comparison is that, in the studied range of modal gain variation $4.4\,\mathrm{cm^{-1}} < \Gamma g_0 < 11.5\,\mathrm{cm^{-1}}$, the impact of filamentation on the lateral beam quality is negligible. This is in contrast to parallel studies of tapered amplifiers, where a similar variation has a large impact on the BPP_{lat} [80].

However, the reduction of the near field modulation could also be beneficial for the broad area laser lifetime, since the local power density at the front facet is reduced and with it the risk of a facet failure. Therefore, a detailed analysis of the near field homogeneity was done. As depicted in figure 3.24(a), the intensity modulation in the near field profile is stronger in the DQW BAL emitter. In order to quantify this observation, the *near field contrast c_{NF}* is introduced as follows:

$$c_{NF} = \frac{max(\text{difference between neighboring intensity minima and maxima})}{\text{mean intensity}}. \quad (3.10)$$

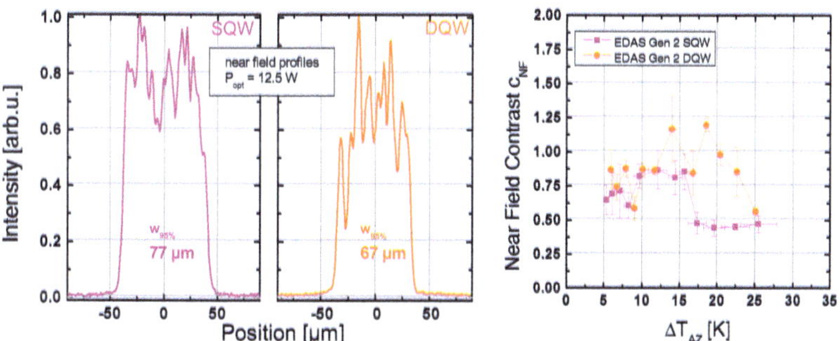

(a) Comparison of near field profiles at $P_{opt} = 12.5\,\mathrm{W}$. (b) Near field contrast c_{NF} as function of ΔT_{AZ}.

Figure 3.24: Analysis of near field homogeneity in EDAS Gen 2 laser diodes. (a) Near field profiles at $P_{opt} = 12.5\,\mathrm{W}$ indicate increased contrast for EDAS DQW design. (b) Averaged near field contrast c_{NF} is decreased by up to 2x for SQW design for $\Delta T_{AZ} > 15\,\mathrm{K}$.

In figure 3.24(b), the averaged near field contrast for both EDAS designs is plotted as function of ΔT_{AZ}. In the temperature regime $\Delta T_{AZ} < 15\,\mathrm{K}$, the two designs show no clear difference, but at $\Delta T_{AZ} > 15\,\mathrm{K}$ ($P_{opt} > 8\,\mathrm{W}$) the near field contrast is strongly reduced in SQW emitters that are predicted to work with reduced filament gain. However, both designs end up with comparable contrast levels at $\Delta T_{AZ} \geq 25\,\mathrm{K}$, so that a clear distinction in terms of contrast is only seen in a limited operation regime: $15\,\mathrm{K} \leq \Delta T_{AZ} \leq 25\,\mathrm{K}$ (contrast-'gap').

In order to assess whether this reduced contrast is beneficial for the BAL reliability, a step-aging test was performed, as shown in figure 3.25. Here, five emitters per species were operated in hard-pulse constant current mode (1s on 1s off, 50 % duty cycle), starting at $P_{opt} = 12\,\mathrm{W}$, followed by a step-wise increase (1 W every 500 hours) up to the maximum comparable power level $P_{max} = 14\,\mathrm{W}$. The power monitoring shows no preference for failures in any of the two groups of emitters.

However, as shown in figure 3.21, the passivation allows output powers up to 19 W and the limit is set by a rapid thermal rollover rather than COMD. Thus, the maximum level of $P_{opt} = 14\,\mathrm{W}$, defined by power saturation effects, might be too low to get a result on improved facet reliability.

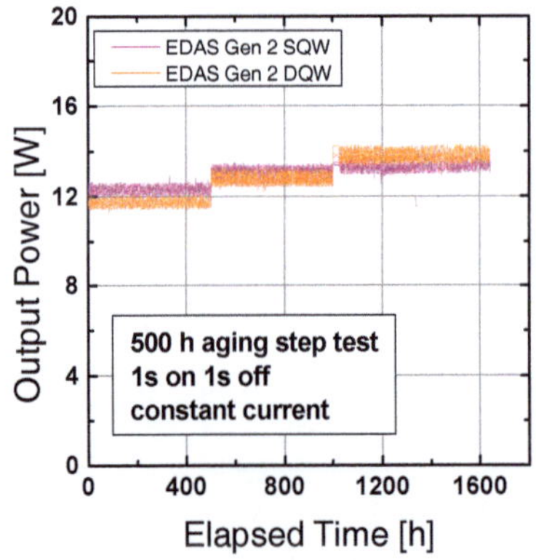

Figure 3.25: Monitored output power in hard-pulse (1s on 1s off) aging step test up to 1600 hours, including 3 steps of power increase and maximum comparable power $P_{opt} = 14\,\mathrm{W}$. No COMD failures were observed and both designs showed stable operation up to peak power.

3.3 Trenches and strain

In this section, the BPP_{lat}-impact of dry etched trenches will be analyzed. The use of etched structures for lateral waveguiding is a well-established technique in the semiconductor laser field [19, 81], especially for ridge-waveguide lasers (RWL). The waveguiding stems from a reduction in the effective refractive index in the etched side regions (quantified via Δn_{eff}), so that guided modes build up in the central part in analogy to a slab waveguide (trench induced 'quasi-index-guiding' [69], see also chapter 2.5 in [18]).

In figure 3.26, the lateral structure of a broad area laser in this study is depicted. The trenches are fabricated using a reactive ion etching (RIE) process, which is strongly anisotropic with an extremely low lateral etch rate, so that the etching results in a rectangular trench profile with steep side walls (cf. chapter 4.1 in [82] and references therein). For optical isolation between neighboring single emitters on a laser bar and for mitigation of lateral lasing, v-grooves are wet-etched through the active area at the laser chip edges. In the central region with the contact opening, the GaAs surface is covered with the p-metal layers. The rest of the semiconductor surface is passivated with a silicon nitride layer.

In this study, the trench depth t and the trench offset (from contact opening) a will be varied. In the following sub-sections, the impact of index guiding trenches on the BPP_{lat}, the degree of polarization ρ and the near field profile will be shown.

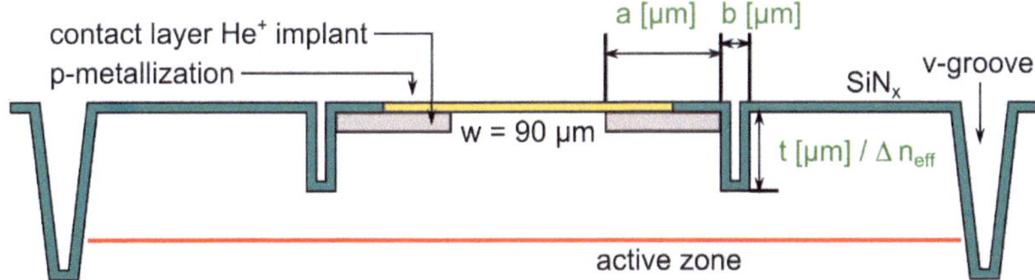

Figure 3.26: Sketch of BAL design with dry etched trenches for lateral index guiding. The contact opening is fixed to $w = 90\,\mu m$ and defined via contact layer implantation with He⁺ ions. The trenches are fabricated using a RIE-process and therefore have a rectangular shape. The etch depth t (and the corresponding effective refractive index step Δn_{eff}), the trench offset from the stripe edge a and the trench width b are important design parameters.

3.3.1 Close trenches

The trench configuration under study has the following properties. The vertical design in use is the EDAS generation one and the chip width is $w_{chip} = 1\,mm$. The trench offset is

$a = 5\,\mu\text{m}$ and the depth (residual layer thickness $d_{res} = 180\,\text{nm}$) is adapted to yield an effective index step of $\Delta n_{eff} = 1.5 \times 10^{-3}$. The trench width is set to $b = 4\,\mu\text{m}$. Since the trench is placed very close to the contact opening, this configuration will be called *'close'* trenches.

In figure 3.27, the near- and far field profiles of two representative BALs are shown for multiple output powers. In the upper row, the index-guided (IG) design with trenches is presented. In the lower row, the gain-guided (GG) design without trenches is shown. From a comparison of figure 3.27(a) with figure 3.27(c), near field stabilization is observed when trenches are used. The IG near fields have steep shoulders and are confined to $|x| \leq 50\,\mu\text{m}$ for all output powers. In contrast to that, the near fields of the gain guided BALs show 'ear-like' side lobes and vary their dimension (measured at 95% power content) at every operating point.

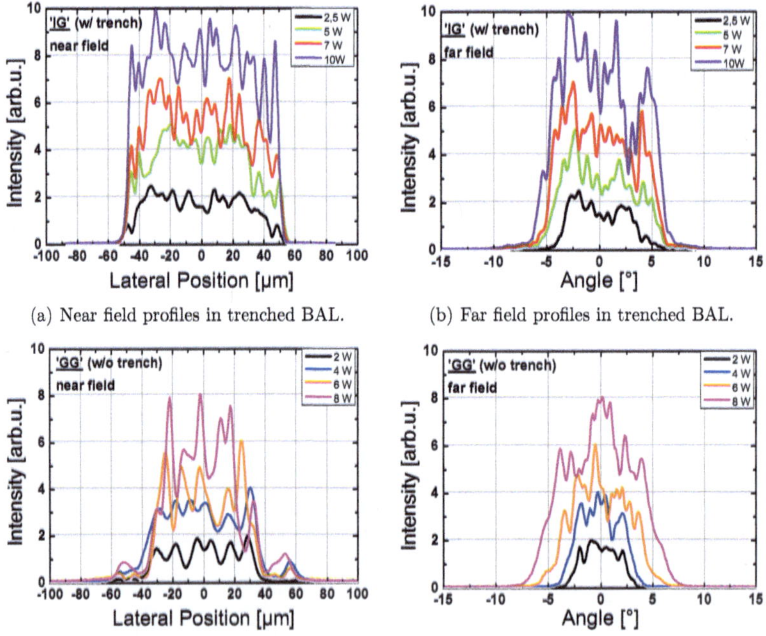

(a) Near field profiles in trenched BAL.

(b) Far field profiles in trenched BAL.

(c) Near field profiles in purely gain guided BAL.

(d) Far field profiles in purely gain guided BAL.

Figure 3.27: Impact of close index guiding trenches on the near- and far field profiles of p-side down mounted EDAS Gen 1 BALs with $w = 90\,\mu\text{m}$ stripe width and $L = 4\,\text{mm}$ resonator length. The field profiles are shown for increasing power output.

The impact of close trenches on the BPP_{lat} is depicted in figure 3.28. As soon as the trenches are etched into the semiconductor, the BPP ground level increases by $\approx$ $1.5\,\text{mm\,mrad}$. Furthermore, a p-up/p-down degeneracy is observed, since for both configurations the linear regression yields almost the same parameters. This result can be explained by the strong optical confinement in the trenched BALs. The field is confined to $|x| \leq 50\,\mu m$ and the difference between the p-side up and p-side down thermal profiles is mainly in the side regions with $|x| > 50\,\mu m$, as seen in figure 3.16(a), where the near field dimensions for the two configurations are indicated as vertical lines. Hence, the difference in the thermal profiles can only affect the BPP_{lat} when the near field expands beyond $|x| = 50\,\mu m$, which is not the case for index guided BALs with close trenches due to the strong lateral near field confinement.

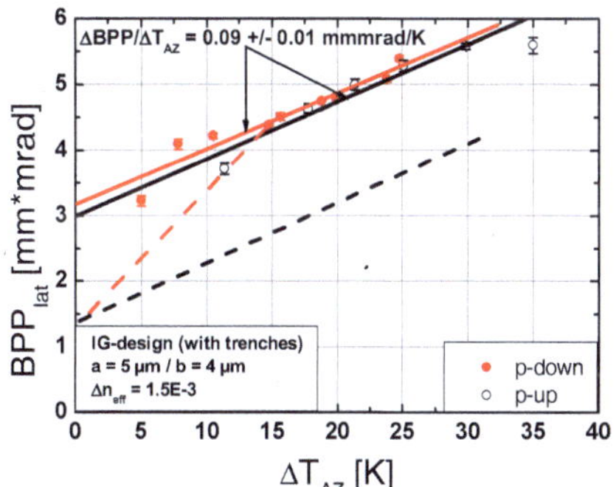

Figure 3.28: BPP_{lat} as function of internal temperature increase for index-guided BALs with close trenches. Solid lines show a linear fit according to equation 3.4 and dashed lines indicate linear fit of gain-guided sibling BALs. Thermal slopes for p-side up (black, empty circles) and p-side down (red, filled circles) mounted BALs are noted.

3.3.2 The impact of strain fields

Another important consideration in diode lasers with close trenches is the impact of strain fields on the BPP_{lat}. As described in [48], the coating of dry etched gallium arsenide edges with anodic oxide can lead to strong strain fields that influence the dielectric function such that effective index differences up to 1×10^{-3} can emerge. Similar effects can be expected here, since in all lateral configurations silicon nitride was used for surface passivation (including the trenches). The strain fields at the emitter edges evoke two concerns. First, the strain causes circular birefringence in the GaAs/AlGaAs crystal [83] that rotates the electric field vector from the transverse electric (TE-) alignment towards transverse magnetic (TM-) alignment, which is perpendicular to the TE-direction. Additionally, hydrostatic pressure also shifts the band edges of the semiconductor [22], increasing gain for TM-radiation as soon as recombination to the light hole valence band is favored. As a consequence, the polarization purity of the laser is reduced, measured via the degree of polarization ρ', as defined in equation 2.18.

The reduction in ρ' is problematic because it makes polarization multiplexing in laser diode modules less efficient. Furthermore, as will be shown below, the TM-part of the laser radiation has a considerably poorer lateral beam quality, so that the overall BPP is increased. Second, the strain-induced refractive index change could introduce additional waveguiding, at worst up to the onset of higher order modes, that causes a BPP_{lat} increase.

In order to analyze the impact of the TM-radiation on the overall BPP, polarization resolved near- and far field measurements were performed. In figure 3.29 and 3.30, the TE- and TM near- and far field profiles for a p-side up mounted and a p-side down mounted BAL with close trenches are shown. The corresponding degrees of polarization $\rho'_{up} = 98\%$ and $\rho'_{down} = 94\%$ show that the p-side down soldering process further decreases the polarization purity by introducing more strain. This assumption is supported by the fact that p-side up mounted, purely gain guided emitters (without trenches) show almost no TM-emission, with $\rho'_{up,notrench} = 99\%$.

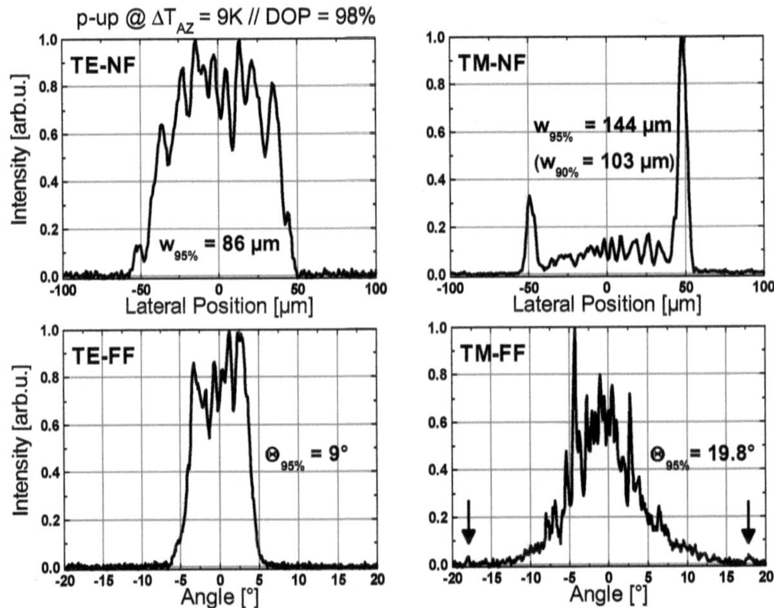

Figure 3.29: Polarization resolved near- and far field profiles of p-side up mounted BAL.

In both mounting types the TM near field shows two spikes at the emitter edges, close to where the trenches are located. This observation is in good agreement with the assumption that the trenches introduce a strain field that triggers circular birefringence and band edge shifts. Hence, the more strain is produced in the stripe edges, the more compromised is ρ' because of a larger amount of TE-radiation that is rotated into TM orientation.

The TM near field spikes broaden the intensity profile and therefore increase the near field waist at 95% power content. However, far more pronounced is the increase in far field divergence for both mounting types. In the p-side up emitter, the far field angle more than doubles from $\Theta_{95\%,TE} = 9°$ to $\Theta_{95\%,TM} = 19.8°$. For the p-side down emitter, a 59% increase is observed from $\Theta_{95\%,TE} = 10.2°$ to $\Theta_{95\%,TM} = 16.2°$. The reason for this strong increase is seen in the TM far field profiles. In addition to a broadened central field, two high-angle side lobes occur at $\pm\ 18°$ (cf. TM-FF in figure 3.30) that increase the 95% power content angle.

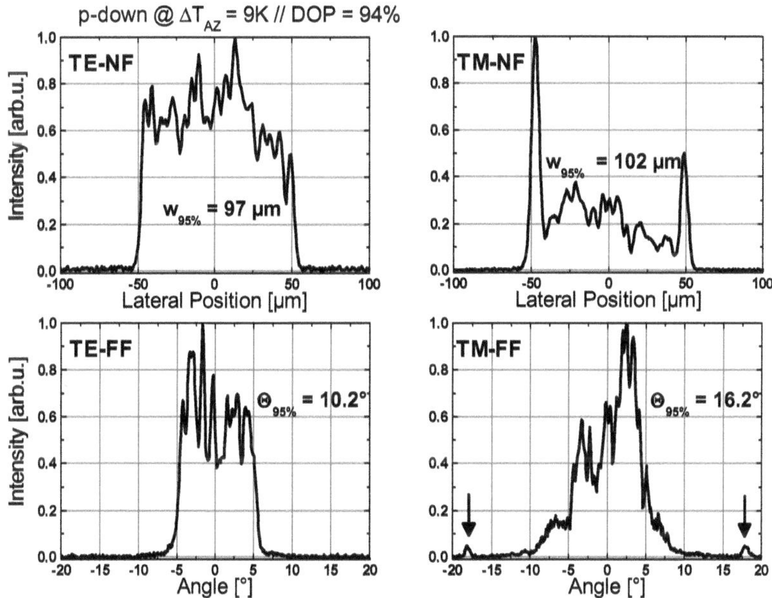

Figure 3.30: Polarization resolved near- and far field profiles of p-side down mounted BAL.

As expected from the shape of the TM-profiles, the $BPP_{lat,TM}$ is strongly increased compared to $BPP_{lat,TE}$, as depicted in figure 3.31 for output powers up to $P_{opt} = 8\,\text{W}$.

In order to infer the overall impact of the high-BPP TM-radiation, a simple superposition estimation was done. The total intensity profile I_{tot} (for both, near- and far field) was generated by artificially varying the degree of polarization:

$$I_{tot}(x) = \rho' \cdot I_{TE}(x) + (1 - \rho') \cdot I_{TM}(x). \tag{3.11}$$

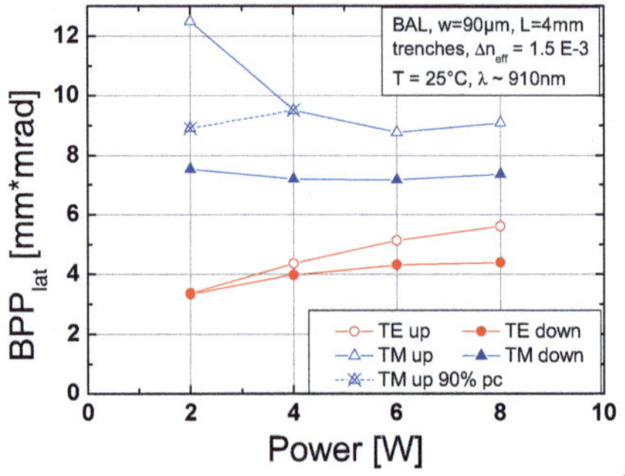

Figure 3.31: Lateral beam parameter product for transverse electric (TE) orientated and transverse magnetic (TM) orientated parts of the BAL laser field as function of output power.

The resulting BPP_{tot} was compared to the pure TE BPP and the difference $\Delta BPP_{lat} = BPP_{tot} - BPP_{TE}$ was calculated. In figure 3.32, the ΔBPP_{lat} is plotted as a function of ρ'.

Even though the TM field of the broad area laser has a strongly degraded beam quality, the overall impact on BPP_{lat} is small as long as the polarization purity is better then 90%. For all devices in this test with close trenches, it was found that $\rho' \geq 94\%$, yielding a maximum BPP penalty of 0.1 mm mrad due to TM radiation with poor beam quality, caused by 'spikes' at the near field edges (below the trenches) and broadened TM far field.

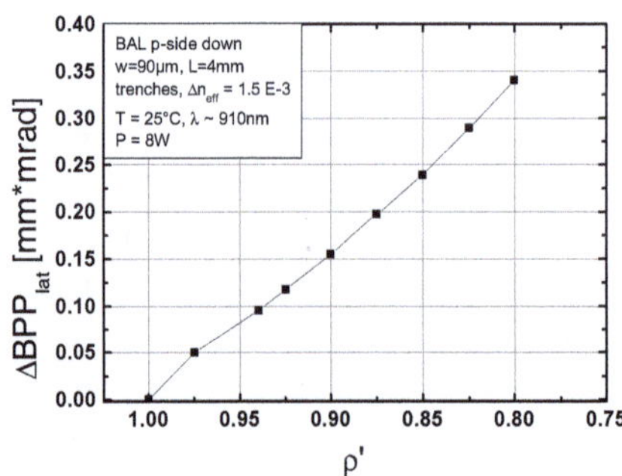

Figure 3.32: Change in lateral beam parameter product due to influence of TM BPP as function of degree of polarization.

3.4 Lateral carrier accumulation (LCA)

Looking back at the listed factors in figure 3.1, the lateral distribution of carriers, which implies the lateral distribution of gain, seems intuitively important. The shape of the lateral gain function determines which modes reach threshold and hence directly influences the BPP_{lat}. The lateral gain distribution has previously been investigated in the literature [68, 84]. Here the focus is set on techniques to control the carrier injection in the active zone. This approach is called 'gain-tailoring'.

In BALs with broad gain regions $w > 30\,\mu m$, the threshold carrier density is first reached in the stripe center, so that only the fundamental mode is amplified just above threshold . With increasing current, the carrier density reaches its threshold value across almost the whole stripe width. However, since the thermal lens constricts all modes in the stripe center, there is an over-pumped edge region left behind. Here, the injected carriers accumulate without being depleted by the existing modes until the gain in this region is high enough to support the next higher order mode. Figure 3.33(a) shows an example of this stripe-edge gain region caused by lateral carrier accumulation (LCA).

In figure 3.33(b), the shape of several higher order modes is depicted. The modes were calculated by solving the 1D Helmholtz equation in a finite difference approach (see section 2.2.2), assuming the thermal lens profile for a p-side down mounted $w = 90\,\mu m$ BAL (cf. section 3.2.1). With increasing mode number, the modes expand in lateral direction, so that the LCA can supply them with gain until they start to lase.

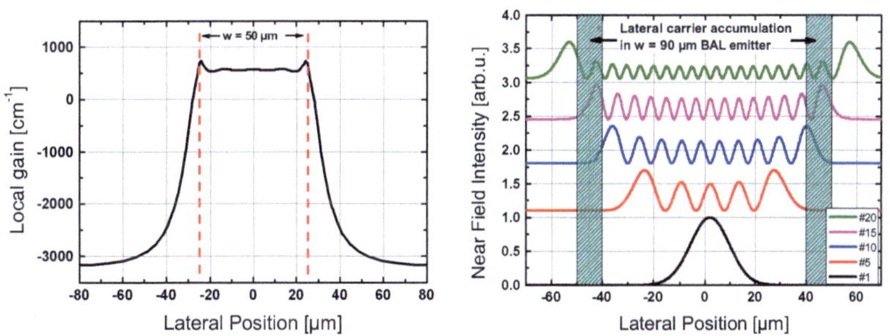

(a) Simulated lateral distribution of gain in a $w = 50\,\mu m$ BAL.

(b) 1D simulation of lateral modes in a BAL.

Figure 3.33: Illustration of lateral carrier accumulation. (a) Simulated lateral gain distribution shows gain increase close to the injection stripe edges. (b) Representative lateral modes in a $w = 90\,\mu m$ BAL, guided by thermal lens taken from figure 3.8(a) ($P_{opt} = 10\,W$). The modes have a hermite-gaussian shape and the extension increases with increasing mode number. The outer maxima overlap with the gain region caused by LCA.

In a recent investigation ([67], based on simulation results), it was stated that a major part (up to 85 %) of the far field blooming in BALs is caused by the non-thermal onset of higher order modes supported by LCA.

Another interesting aspect is the high gain-loss contrast between the stripe region and the surrounding areas. They are pumped with carriers via lateral carrier escape along the QW and lateral current spreading. Although they do not contribute to the stimulated emission, they cause strong free carrier absorption losses. In purely gain-guided devices, this gain loss contrast is the main guiding mechanism in the lateral direction, besides the thermal lens. It is known that this guiding via the imaginary part of the refractive index causes a phase front bending [85], which influences the far field pattern. However, it is not clear how pronounced this effect is in BALs and by how much BPP_{lat} is changed.

In order to analyze the impact of the LCA on BPP_{lat} in BALs, a series of test devices with LCA-suppression was produced and assessed in this study. Here, the mitigation of LCA was achieved via deep proton implantation [7], which will be explained in the following subsection.

3.4.1 Deep proton implantation

Implantation is a well established technique in GaAs laser processing for achieving a well defined current injection without etching of the semiconductor or the deposition of insulators [54, 56]. Usually, the ions of light elements (hydrogen H^+ or helium He^+) are accelerated in a strong electric field (kV-range) before they are shot into the semiconductor crystal. There they collide with atoms of the crystal lattice and produce crystal defects, as described in section 2.2.3. As a consequence the electrical and optical properties of the bombarded material change. The high density of defects decreases the electrical conductivity of the crystal lattice and increases the optical absorption. However, as explained in [57] a temperature annealing can reverse the optical damage while the electrical conductance stays low.

In this study, proton bombardment was used to implant the regions beyond the injection stripe, as shown in figure 3.34. The implantation was applied to EDAS Gen 2 BAL emitters with $w = 90\,\mu m$ and consisted of two steps. First, a dose of 1×10^{14} cm^{-2} at an energy of $E_{H+} = 230\,keV$ and then a second dose of 1.4×10^{14} cm^{-2} at an energy of $E_{H+} = 340\,keV$. All devices received an annealing at $T = 300\,°C$ for one hour, which is sufficient to eliminate implantation-induced optical losses [57]. The expected implantation depth is $d_{implant} = 3.9\,\mu m$, calculated with Monte-Carlo simulation (SRIM [53]), such that the n-side of the device is reached. The defect density in the active region is $\approx 10^{18}$ cm^{-3}.

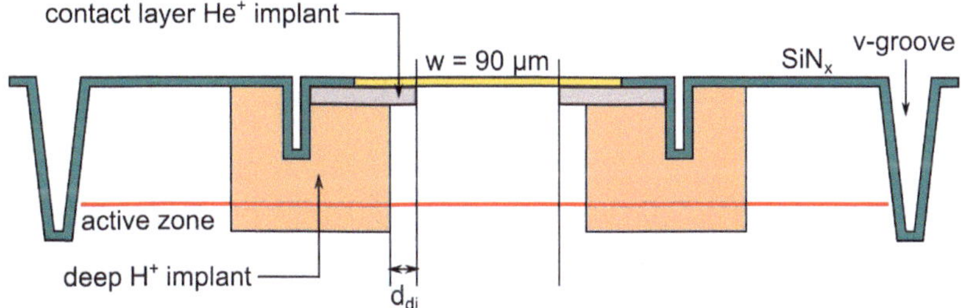

Figure 3.34: Lateral design of deeply proton-implanted BALs. The implantation depth is sufficient to penetrate the quantum well and reach the n-side of the diode. The implantation offset was varied: $d_{di} \in \{0, 5, 10\}$ µm.

With this implantation design, the current spreading above the QW is blocked by the induced regions of high resistivity and the defect centers in the QW-edges lead to the depletion of accumulated carriers via rapid non-thermal recombination. As a result, the gain loss contrast at the stripe edges is minimized. As a diagnostic tool, the lateral distance d_{di} of the implantation mask was varied (cf. figure 3.34).

Verification of implantation goal

In order to verify that the implantation process was effective, a series of diagnostic experiments was done. First, the implantation depth and the implantation width were measured in an SEM, using cathodoluminescence (CL, cf. figure 3.35). The top view via the etched substrate of a non-implanted BAL in figure 3.35(a) shows radiative recombination of injected carriers in the whole laser chip. In contrast, the implanted BAL with $d_{di} = 0$ µm is shown in figure 3.35(b). Here the measured stripe width is $w = 89$ µm, which is very close to the intended width ($w = 90$ µm), such that the implantation does not spread laterally into the pumped region. In the cross-sectional CL image in figure 3.35(c), the implanted regions show up as dark areas in the n-side. In the p-part of the diode, the CL does not produce any contrast, since here carbon is used as dopant [86]. The implantation depth was determined as $d_{implant} = 3.9$ µm, which is consistent with the Monte-Carlo simulation.

In a second assessment, it was tested if the implantation influenced the current spreading. The first positive indicator is the series resistance R_s, which increases by 12.2 % from 18.9 mΩ to 21.2 mΩ for BALs with $d_{di} = 0$ µm compared to non-implanted emitters. Simultaneously the threshold current decreases by 12.2%, from 1183 mA to 1039 mA. This mirrors a narrowed cross-sectional area where the current can pass. However, to get a more detailed picture two more tests were performed. First, the amplified spontaneous emission (ASE) near field was measured below threshold for different implantation distances.

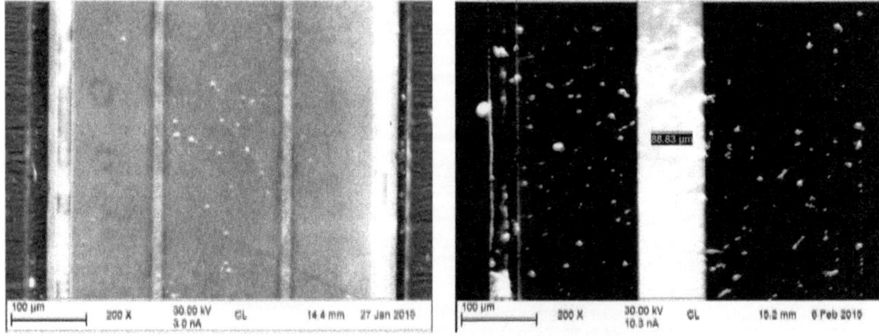

(a) Top view CL of non-implanted BAL.   (b) Top view CL of implanted BAL.

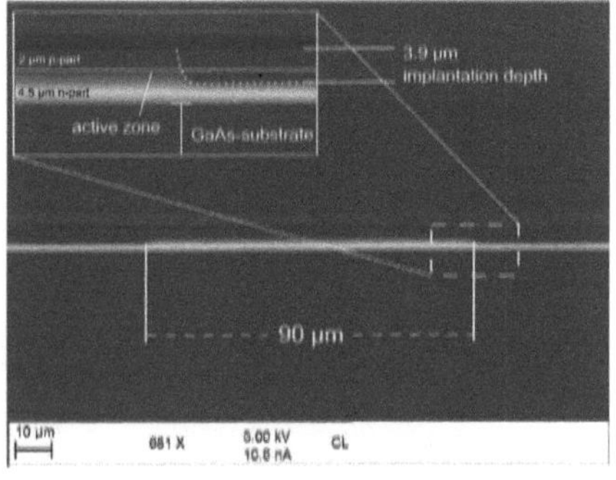

(c) Lateral cross section CL.

Figure 3.35: Cathodoluminescence (CL) measurements. (a) Plan view CL-image of non-implanted BAL shows luminescence in the whole laser chip. (b) The deeply implanted BAL with $d_{di} = 0\,\mu m$ shows successful suppression of radiative carrier recombination in the device edges. (c) Lateral cross section shows implantation depth of $d_{implant} = 3.9\,\mu m$ is reached as expected from calculation.

As shown in figure 3.36(a), the near field width decreases continuously with decreasing d_{di}: $w_{95\%} = 198\,\mu m$ for non-implanted BALs, $w_{95\%} = 153\,\mu m$ for $d_{di} = 10\,\mu m$ and $w_{95\%} = 140\,\mu m$ for $d_{di} = 0\,\mu m$. This result shows that less radiative recombination is observed in the edge regions of the stripe and indicates successful lateral current confinement as soon as the implantation gets close to the stripe edge.

The second experiment is the measurement of the carrier non-clamping in the active region at the device edges, using a fiber as emission collector, as shown schematically in figure 3.36(b) and first proposed by Smowton and Blood [87]. Here, the spontaneous emission from the front facet is collected with a fiber that has a $\theta = 45°$ angle to the optical axis to avoid stimulated emission input. In theory the spontaneous emission increases as the current increases up to the threshold and is constant for $I > I_{thr}$ because the carrier density in the active region is pinned at the threshold level. However, due to lateral current spreading, carriers are injected into the lateral side regions beyond the stripe edges where they reach the QW and recombine. Since the carrier density in these regions has not reached the transparency level, a further increase in spontaneous emission is observed and its slope correlates with the amount of lateral carrier spreading.

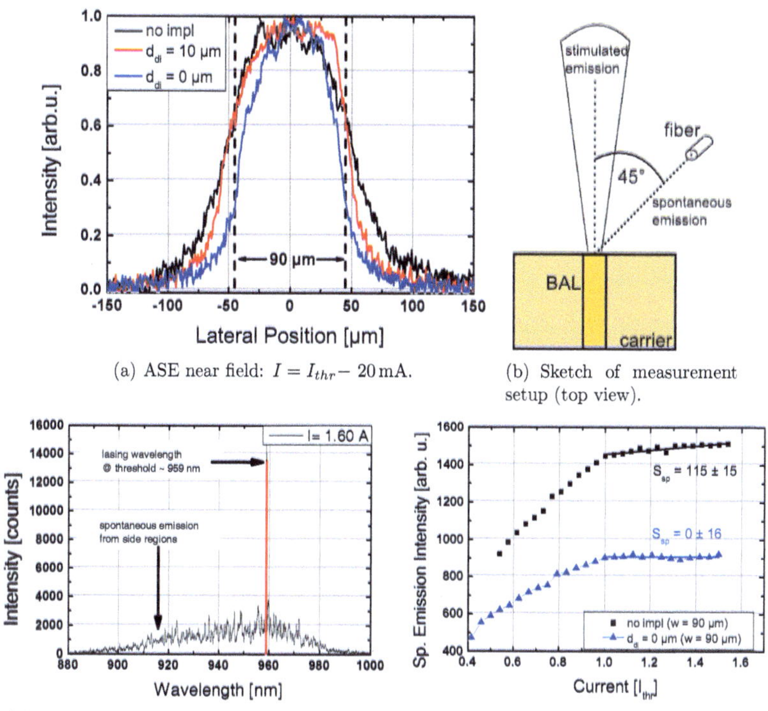

(a) ASE near field: $I = I_{thr} - 20\,\text{mA}$.

(b) Sketch of measurement setup (top view).

(c) Spectrum of high-angle spontaneous emission.

(d) Spontaneous emission intensity slope.

Figure 3.36: Verification of current spreading suppression. (a) ASE near field profiles of implanted emitters with different d_{di}. (b) Schematic sketch of measurement setup. A fiber collects the spontaneous emission at an angle of $\theta = 45°$. (c) The intensity is integrated for $\lambda \leq \lambda_{st.em.}$ to exclude photons produced by stimulated emission. (d) In implanted emitters, the spontaneous emission intensity is pinned above threshold, so that lateral carrier escape is successfully suppressed.

In this experiment, the slope of the spontaneous emission intensity with current S_{sp} is used to quantify the level of current spreading in non-implanted and deeply proton-bombarded emitters. If the current spreading is successfully mitigated with implantation, S_{sp} should be reduced. In figure 3.36(c), a representative spontaneous emission spectrum is shown. The intensity for the wavelengths $\lambda \leq \lambda_{st.em.}$ was integrated for every current value below and above threshold, so that stimulated emission is excluded. The comparison in figure 3.36(d) between the non-implanted and the implanted BAL with $d_{di} = 0\,\mu\text{m}$ shows a clear reduction in S_{sp} above threshold from (115 ± 15) to (0 ± 16) counts per mA. The latter slope even indicates a complete pinning of the spontaneous emission above threshold. However, due to the measurement uncertainty, this statement cannot be fully proved.

In summary, these diagnostic experiments show that deep proton implantation is a precise and predictable technique to control the current flow in GaAs broad area lasers. Furthermore, it was confirmed that implantation has a strong impact on the lateral carrier profile at low bias (CL), just below threshold (ASE-measurement) and above threshold (spontaneous emission from emitter edges).

3.4.2 The effect of reduced LCA

The LIV curves in figure 3.37(a) show that implantation at a distance $d_{di} = 10\,\mu\text{m}$ has almost no impact on the slope S. In contrast to this, the very close implantation at $d_{di} = 0\,\mu\text{m}$ reduces the slope and causes the power-curve to rollover early. This impression is confirmed by the efficiency curves in figure 3.37(b). The LIV results are summarized in table 3.4.

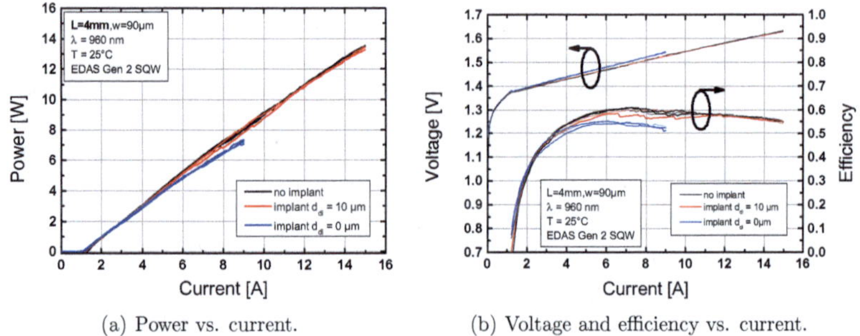

(a) Power vs. current.

(b) Voltage and efficiency vs. current.

Figure 3.37: Basic LIV characterization of deeply implanted BAL emitters. (a) Output power P_{opt} as function of injection current. For $d_{di} = 10\,\mu\text{m}$, the power curves are identical to that of non-implanted emitters. For $d_{di} = 0\,\mu\text{m}$, a reduced slope is observed. (b) Voltage and efficiency as function of current. Emitters with $d_{di} = 0\,\mu\text{m}$ show reduced conversion efficiency.

parameter	unit	no implant	$\mathbf{d_{di}} = 10\,\mu\mathrm{m}$	$\mathbf{d_{di}} = 0\,\mu\mathrm{m}$
I_{thr}	[mA]	1180	1084	1040
S	[W A^{-1}]	1.04	1.04	0.98
$\eta_{c,max}$	[%]	60.7	60.4	55
R_s	[mΩ]	18.9	18.7	21.2

Table 3.4: Summarized results from LIV-characterization of deeply implanted emitters with varying d_{di}.

As the slope S, the peak conversion efficiency $\eta_{c,max}$ and the series resistance R_s are almost unchanged with $d_{di} = 10\,\mu\mathrm{m}$ implantation, the critical region for lateral carrier effects is contained within $10\,\mu\mathrm{m}$ beyond the injection stripe edge. Interestingly, the reduction in threshold current I_{thr} is already observed at $d_{di} = 10\,\mu\mathrm{m}$, with only minimal further decrease for $d_{di} = 0\,\mu\mathrm{m}$. In figure 3.38, the impact of implantation on near- and far field is depicted. For emitters with $d_{di} = 10\,\mu\mathrm{m}$, the near field width as function of current is comparable to the non-implanted near field behavior and the far field blooming is only slightly reduced. In contrast, the emitters with $d_{di} = 0\,\mu\mathrm{m}$ show a significantly reduced near field width *and* a reduced far field divergence for currents up to $I = 8.5\,\mathrm{A}$.

The reduced near- and far field dimensions for $d_{di} = 0\,\mu\mathrm{m}$ cause a reduction in BPP_{lat}, as shown in figure 3.39. At $P_{opt} = 7\,\mathrm{W}$ (highest comparable power level), a decrease by 23% from $2.6\,\mathrm{mm\,mrad}$ to $2\,\mathrm{mm\,mrad}$ is observed. For $d_{di} = 10\,\mu\mathrm{m}$, no clear improvement in terms of beam quality is seen. The analysis in figure 3.39(b) shows a decreased thermal slope S_{th} for $d_{di} = 0\,\mu\mathrm{m}$ with $S_{th} = (0.11 \pm 0.01)\,\mathrm{mm\,mrad\,K^{-1}}$, which is a 35%-reduction compared to non-implanted emitters with $S_{th} = (0.17 \pm 0.01)\,\mathrm{mm\,mrad\,K^{-1}}$. All beam quality results are summarized in table 3.5.

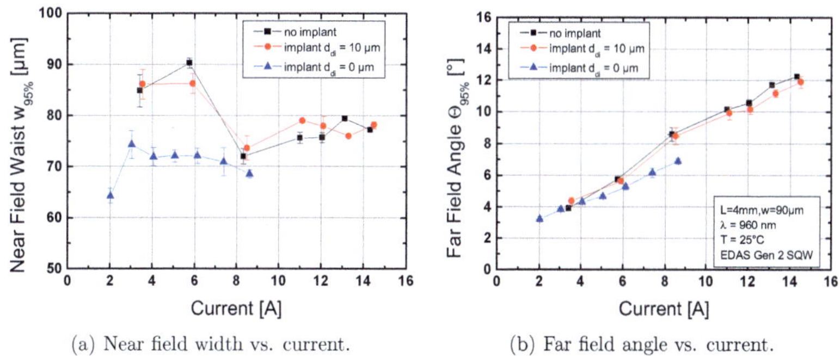

(a) Near field width vs. current.

(b) Far field angle vs. current.

Figure 3.38: Near field width and far field angle at 95% power content as function of current for deeply implanted BALs.

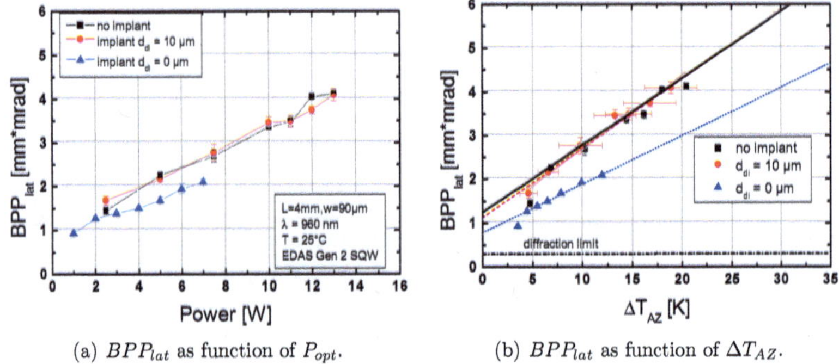

(a) BPP_{lat} as function of P_{opt}.

(b) BPP_{lat} as function of ΔT_{AZ}.

Figure 3.39: Lateral beam parameter product BPP_{lat} as function of power and active zone temperature increase ΔT_{AZ} for deeply implanted BALs.

In summary, the deep proton implantation only affects the BAL performance if it is close enough (i.e. $d_{di} < 10\,\mu m$) to the injection stripe edge. A very close implantation at $d_{di} = 0\,\mu m$ yields improved beam quality, but compromises the conversion efficiency. This confirms that the region of $\pm\,10\,\mu m$ from the stripe edge is critical for carrier profile manipulation. According to the model in [88], the internal differential efficiency can be written as a product of three sub-efficiencies:

$$\eta_o^d = \eta_s^d \cdot \eta_i^d \cdot \eta_r^d, \tag{3.12}$$

parameter	unit	no implant	$d_{di} = 10\,\mu m$	$d_{di} = 0\,\mu m$
S_{th}	[mm mrad K^{-1}]	0.17 ± 0.01	0.16 ± 0.01	0.11 ± 0.01
BPP_0	[mm mrad]	0.9 ± 0.3	1.0 ± 0.1	0.75 ± 0.03
BPP_{lat} (7 W)	[mm mrad]	2.6	2.6	2
$w_{95\%}$ (7 W)	[μm]	72 ± 2	74 ± 2	69 ± 1
$\Theta_{95\%}$ (7 W)	[°]	8.6 ± 0.4	8.5 ± 0.5	6.9 ± 0.2

Table 3.5: Summarized results from beam quality measurement of deeply implanted emitters with varying d_{di}.

where η_o^d is the overall internal differential efficiency. The *current spreading* efficiency η_s^d accounts for the fraction of the current that enters the active geometrical area where the round-trip gain is sufficient to match the optical losses. The quantum well *injection* efficiency η_i^d represents the fraction that results in an increase in quantum well recombination current. It is decreased by recombination in barrier and waveguide and drift and diffusion through the cladding layers. Finally, the *radiative* efficiency η_r^d represents the fraction of carriers that reach the quantum well and recombine radiatively.

Now, in deeply implanted emitters the following hypothesis is assumed. Due to the blocking of lateral current flow, η_s^d is expected to be close to unity. However, since the defect density beyond the stripe edge is very high in the active zone the other two factors must have been affected negatively. Most probably η_r^d decreased substantially, since the defect centers allow rapid non-radiative carrier recombination.

Overall, the result of this study is that lateral carrier accumulation affects the BPP_{lat} significantly, since 33% of the thermal slope S_{th} is caused by this phenomenon. However, due to substantial efficiency loss, the methods for reducing LCA must be improved. This could be done by adapting the implantation recipe or by a completely different approach to lateral structurization, such as quantum well intermixing or etch and re-growth techniques.

Aging test of deeply implanted emitters

In order to obtain information on the long-term stability of the proton implantation effect, a sample group of implanted BALs was tested in an aging experiment with the following conditions. Two emitters per species (no implantation, $d_{di} = 10\,\mu\text{m}$ and $d_{di} = 0\,\mu\text{m}$) were tested in constant current ($I = 8\,\text{A}$) hard-pulse (1 s on 1 s off) mode for a total of 5000 h. Due to a different position of the temperature sensor in the aging submount-holders, the operating temperature was adapted to $T_{HS} = 15\,°\text{C}$ in order to maintain the junction temperature used previously in the characterization measurements. In figure 3.40, the monitored output power during the aging is shown. Across the whole time span, only slight deviations are observed and no emitter failed during the test.

After 1000 h, 3000 h and 5000 h, all emitters were re-tested in LIV, spectrum and beam quality measurements. For convenience, only the performance comparison after 5000 h will be presented in the following. In figure 3.41, the impacts of aging on optical power output P_{opt}, conversion efficiency η_c and voltage are shown. Apart from a slight deviation in the power output for $P_{opt} > 5\,\text{W}$, no significant changes were observed.

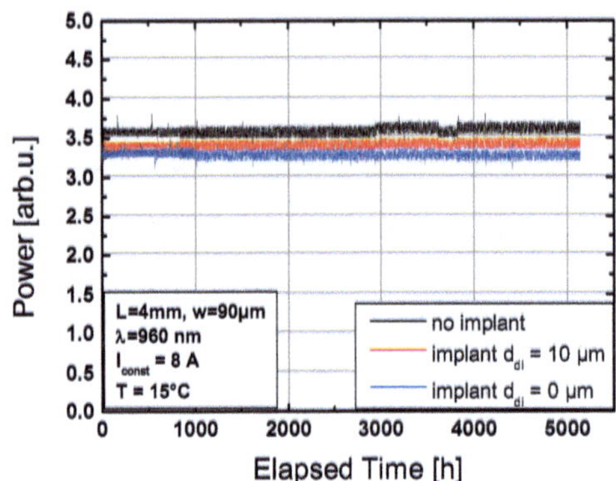

Figure 3.40: 5000 h hard-pulse mode aging test for deeply implanted BALs and non-implanted reference.

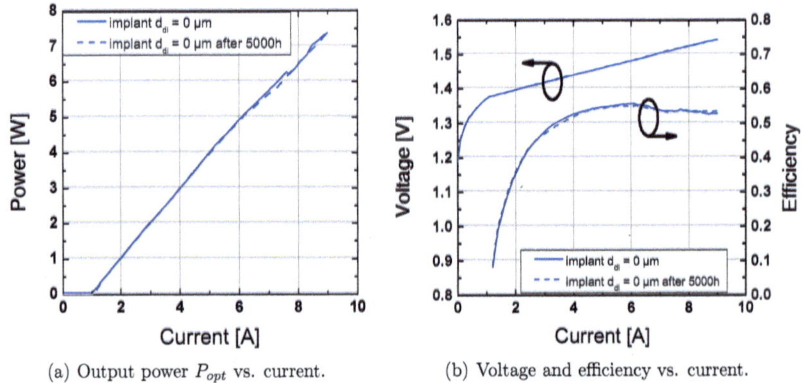

(a) Output power P_{opt} vs. current.   (b) Voltage and efficiency vs. current.

Figure 3.41: Comparison of BAL performance before and after 5000 hours of aging. (a) Output power curve $P_{opt}(I)$ shows only slight deviations after aging. (b) Voltage- and efficiency-curves remain unchanged.

The near field widths after aging in figure 3.42(a) show a different progression. However, for every operating point, the difference to the original width lies within the error-range. The mean far field divergence is slightly increased, as shown in figure 3.42(b). The difference to the original angles peaks at $P_{opt} = 5\,\text{W}$ with $\Delta\Theta_{95\%} = 0.6°$. However, the deviation is not observed in all emitters, as will be shown in figure 3.43.

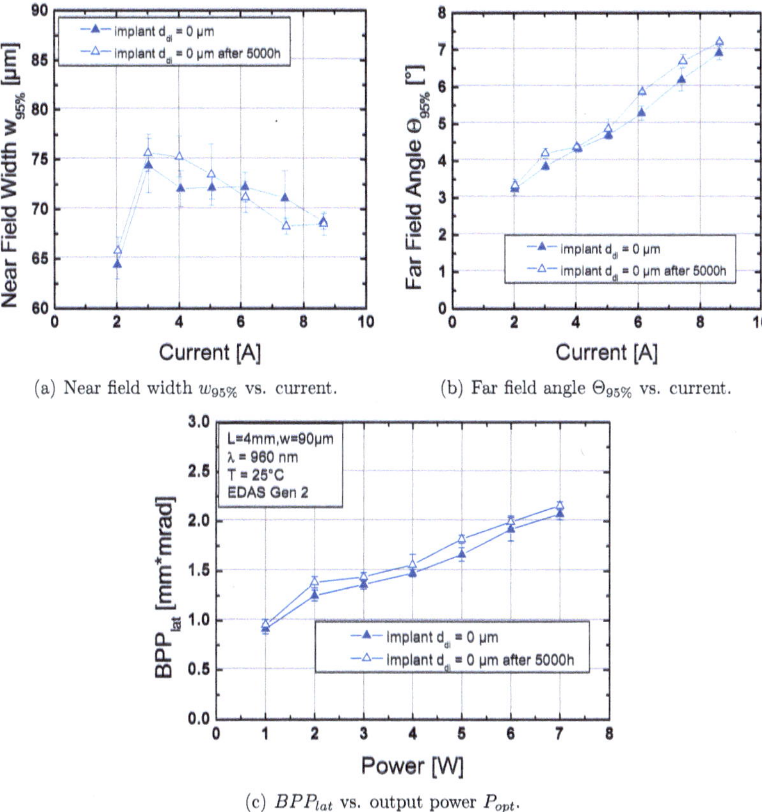

(a) Near field width $w_{95\%}$ vs. current.

(b) Far field angle $\Theta_{95\%}$ vs. current.

(c) BPP_{lat} vs. output power P_{opt}.

Figure 3.42: Aging impact on beam quality in deeply implanted BALs after 5000 hours. (a) Near field width and (b) Far field angle as function of current. The far field divergence is slightly increased. (c) Lateral beam parameter product increase is at most 0.2 mm mrad.

The increased far field angle after $5000\,\text{h}$ yields an increased BPP_{lat} as well (cf. figure 3.42(c)). Here, the most extreme deviation is a difference of $\Delta BPP_{lat} = 0.2\,\text{mm mrad}$. However, as mentioned above, some emitters show only slight alterations of the field profiles. In figure 3.43, the near- and far field profiles of an exemplary BAL emitter with $d_{di} = 0\,\mu\text{m}$ are shown before and after aging. In both fields only minimal changes are noted after the $5000\,\text{h}$ aging test. However, as noted in figure 3.43(a), the near field widths after 1000 h and 3000 h differ only slightly ($\Delta w_{95\%} \leq \pm 2\,\mu m$) from the value after 5000 h aging.

In summary, the aging test shows that the improved beam quality properties achieved via deep proton implantation are stable after 5000 h of hard pulse operation. No emitter-failure was observed and deviations in BPP_{lat} lie within an acceptable range. At $P_{opt} = 7\,W$, the BPP_{lat} changes by 0.08 mm mrad from 2.06 mm mrad to 2.14 mm mrad, yielding $\approx 4\,\%$ deviation, which is within the measurement uncertainty of $\pm 5\,\%$ (assuming BPP_{lat} measurement accuracy of ± 0.1 mm mrad, cf. section 2.3).

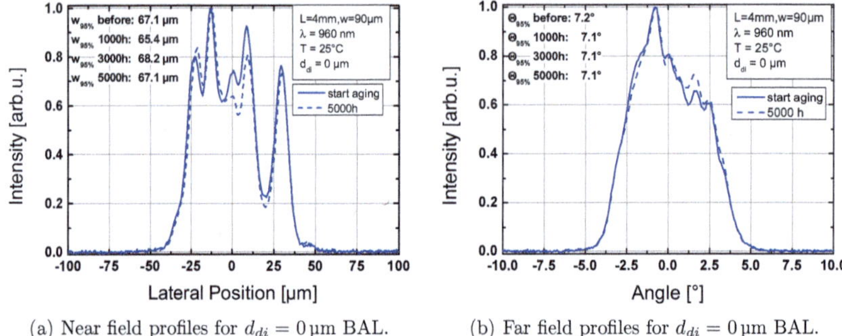

(a) Near field profiles for $d_{di} = 0\,\mu m$ BAL. (b) Far field profiles for $d_{di} = 0\,\mu m$ BAL.

Figure 3.43: Near- and far field profiles of deeply implanted BAL with $d_{di} = 0\,\mu m$ before and after 5000 hours of hard pulse aging test. (a) Near field intensity shows stable profiles during aging and the near field width changes slightly. (b) The far field profile retains its shape after 5000 hours of aging.

3.5 Summarized lessons and measures for improved BPP_{lat}

Looking back at the list of possible influences in the root cause diagram (cf. figure 3.1), the following insights were revealed during this study.

Lateral profile of thermal lens and vertical design

A strong correlation between thermal lens bowing B_2 and the thermal BPP_{lat} deterioration rate S_{th} was observed. An increasingly bowed thermal lens leads to a faster BPP_{lat} increase. It should be noted here that the *absolute* bowing values determined with micro-thermography are distorted by a broadening effect and are inappropriate for absolute quantification.

Two methods to modify the thermal lens profile were found in this study. First, the use of an epitaxial layer structure with strong asymmetry (EDAS) resulted in a stronger thermal lens bowing as compared to a conventional, less asymmetric (ASLOC) design. The latter showed a 50% reduction in S_{th}, but a simultaneous 50% increase in BPP_0. The use of a p-side GRIN layer close to the active region and the p-layer aluminum content, which influences the thermal conductivity, are suspected to play a major role in shaping the thermal lens. Second, a reduction in the size of the laser chip, i.e. 42 % increase in d_{p-side} and 50% chip width reduction, yielded a 33% decrease in S_{th}. A decrease in BPP_0 is observed as well and the amount depends on the active region. For emitters with InGaAs DQW and GaAsP barriers a 15% reduction is observed and for a SQW InGaAs active region with GaAs spacer the drop in BPP_0 increases to 39%.

A two-dimensional waveguide mode simulation was performed in order to assess the effect of asymmetry (EDAS vs. ASLOC) on the confinement and the gain of individual modes. The strongly asymmetric EDAS design showed a marginally faster decrease of the relative mode gain and hence has a stronger suppression of higher order modes (assuming equal thermal waveguides for both designs). This property of the EDAS design is expected to favor the reduction in the BPP_0 level, compared to the more symmetric ASLOC design.

Filamentation

On basis of theoretical calculations [77] that predict a lower filament gain γ_{max} in broad area lasers with reduced differential gain $\Delta G/\Delta j$, two structures with different Γg_0 (but otherwise identical) were tested in terms of beam quality. The decrease in modal gain from $\Gamma g_0 = 11.5\,\mathrm{cm}^{-1}$ to $4.4\,\mathrm{cm}^{-1}$ yielded a -54% reduction in γ_{max}. The low γ_{max} devices showed a -48% smaller near field contrast (at $P_{opt} = 10\,\mathrm{W}$) but no considerable impact on BPP_{lat} was observed. Therefore, it is concluded that the effect described in [77] is already saturated at $\Gamma g_0 = 12\,\mathrm{cm}^{-1}$ and a further modal gain decrease is not beneficial for BALs.

Index guiding trenches

The use of 'close' index guiding trenches (trench offset $a = 5\,\mu\mathrm{m}$, trench width $b = 4\,\mu\mathrm{m}$ and effective index step $\Delta n_{eff} = 1.5 \times 10^{-3}$) resulted in a moderate thermal slope of $S_{th} = 0.09\,\mathrm{mm\,mrad\,K}^{-1}$ and a BPP_0 increase by $1.5\,\mathrm{mm\,mrad}$. The strong trench induced waveguiding causes stabilized near field profiles with steep side shoulders for all output powers. Furthermore, it suppressed other waveguiding mechanisms, so that the $BPP_{lat}(\Delta T_{AZ})$ curves for p-side down and p-side up mounted BALs are the same.

Mechanical strain

The beam quality comparison of BALs with and without (close) index guiding trenches led to the following insights. The use of dry etched trenches that are covered with silicon nitride introduces material strain that causes a reduction in the degree of polarization ρ' of up to 6%-points in the most extreme configuration (trenches and p-side down mounting). A detailed evaluation of the polarization resolved near and far field profiles showed that trenched BALs have considerably poorer TM BPP_{lat}, which is a factor of $\approx 1.7\text{x}$ larger than TE BPP_{lat}. This increase is caused by TM near field peaks at the stripe edges (where the trenches are located) and an increased TM far field divergence. However, it was shown that the impact on the total BPP_{lat} stays below $0.1\,\text{mm}\,\text{mrad}$ as soon as $\rho' > 94\%$.

Lateral carrier profile and proton implantation

In order to assess the impact of lateral carrier accumulation (LCA) on the BPP_{lat} in broad area lasers, a diagnostic study was performed. The LCA was suppressed by using deep proton implantation at the emitter stripe edges, that reaches the n-side of the diode laser and penetrates the quantum well. Two effects helped to achieve this. First the lateral current spreading was prevented due to the reduced electrical conductivity in the proton-bombarded GaAs. And secondly, the implantation induced defect density of $1 \times 10^{18}\,\text{cm}^{-3}$ in the quantum well enables rapid non-radiative recombination of carriers at the device edges, such that accumulation is prevented.

The characterization of implanted emitters showed that the BPP_{lat} is influenced if the implantation gets very close to the stripe edge ($d_{di} = 0\,\mu\text{m}$). At $P_{opt} = 7\,\text{W}$ the BPP_{lat} decreased from $2.6\,\text{mm}\,\text{mrad}$ for non-implanted BALs to $2\,\text{mm}\,\text{mrad}$, yielding a linear radiance of $B_{lin} = 3.5\,\text{W}(\text{mm}\,\text{mrad})^{-1}$, which is comparable to best reported industry standards (cf. table 1.2 in section 1.1). The thermal slope S_{th} decreased by 33% in BALs with $d_{di} = 0\,\mu\text{m}$ implantation, which is consistent with theoretical predictions. However, the LCA-effect is current driven and the reduced thermal slope is expected to stem from the current dependency of $\Delta T_{AZ} \propto I$. The BPP improvement is accompanied by a reduction in conversion efficiency η_c which shrinks by 7% due to the high defect density at the emitter edges.

Since no effect on BPP_{lat} is observed for $d_{di} = 10\,\mu\text{m}$, it is assumed that the $10\,\mu\text{m}$ regions beside the injection stripe are critical for the manipulation of the lateral carrier distribution.

Measures for improved BPP_{lat}

In summary, three design rules are deduced from this study. First, the thermal lens profile should be smoothed in order to decrease the thermal BPP_{lat} deterioration rate S_{th}, since this parameter is directly coupled to the thermal lens bowing. In the literature, a couple of thermal lens shaping attempts were published [89–92]. However, the boundary condition is always to improve the thermal lens shape *without* compromising efficiency η_c and degree of polarization ρ'. Further investigation is needed to clarify the impact of the epitaxial design (especially the p-side of the structure), the chip-geometry and also the p-side metallization on the thermal lens shape.

Secondly, a strong index-guiding trench close to the stripe should be avoided. It introduces TM-radiation with poor beam quality and increases the BPP ground level BPP_0. However, index-guiding also has the advantage of near field stabilization and moderate thermal slope S_{th}. There is a trade-off between the positive stabilization effect and the BPP_0 increase. The most important parameters here are trench-offset and the effective refractive index step Δn_{eff}. These two should be optimized in a test-matrix to yield the highest linear radiance for the broad area emitter.

Last, but not least, the suppression of lateral carrier accumulation is helpful in decreasing BPP_{lat}. The deep proton implantation approach used here was helpful in terms of diagnostic insight. A zone of $\approx \pm 10\,\mu m$ at the stripe edges was identified to be critical for LCA suppression. However, for high power BALs with high efficiency, other methods to reduce LCA must be sought since the damaged quantum well caused an efficiency penalty. Here, another trade-off design might be possible that works with adapted dose and implantation depth in order to minimize the efficiency loss and maintain as much BPP_{lat} improvement as possible. Alternative techniques are quantum well intermixing [93] and etch-and-regrowth techniques (both rely on an increased bandgap in the emitter edges to push carriers back into the stripe center and avoid photon absorption).

Chapter 4

Further Measures for Improved Slow Axis Beam Quality

In the previous chapter, the focus was set on a root cause analysis of BPP_{lat} deterioration in BALs with fixed geometry and test conditions ($w = 90\,\mu m$ and $L = 4\,mm$). By using diagnostic experiments that isolate potentially important device parameters, the knowledge about the main influences on BPP_{lat} was extended. This chapter, however, deals with measures that aim to reduce BPP_{lat} directly by reducing the number of active lateral modes. It is well known that multi-mode operation in BALs compromises the beam-quality [40, 46]. The occurrence of multiple lateral modes is due to the thermal lens, which induces a thermal waveguide, and the broad, laterally uniform gain profile that does not provide any intrinsic mode discrimination. On the contrary: Higher order modes extract the lateral gain profile more effectively than the fundamental mode [46].

In the following, two BAL design measures that seek to reduce the number of lateral modes are investigated. First, a purely geometrical approach: In order to find an optimum in the trade-off between beam quality improvement and efficiency decrease, a series of BALs with varied stripe width w is assessed. Secondly, a lateral mode filter approach is presented that uses a high-optical-loss vertical waveguide beside the laser waveguide. This approach is based on the adaptation of the effective refractive indices n_{eff} in the neighboring vertical waveguides and filters higher order modes due to their greater near field extent $w_{95\%,m} > w_{95\%,n}$ for $m > n$, with m, n being lateral mode indices.

4.1 Stripe width variation

In diode lasers with strong index guiding (e.g. through etch and regrowth or ridge waveguides), the lateral modes can be calculated with the effective index method [94]. In this case, the number of lateral modes q can be estimated via the normalized waveguide width $W \propto w$, such that $q \cong W/\pi$ (cf. chapter 2.5 in [18]), which directly shows that fewer modes are present in emitters with smaller stripe widths w.

In BALs with pure gain guiding or weak index guiding trenches (i.e. $\Delta n_{eff} < 1\text{E-}4$, 'quasi-index-guiding' [69]), the situation is different. Here, the thermal waveguide that builds up during operation is dominant. In this section, it will be shown that for equal ΔT_{AZ}, a thermal waveguide generated in a narrow stripe guides fewer modes than thermal waveguides in broad stripes. Thus, it still holds that q grows with increasing w and hence BPP_{lat} is expected to be lower for reduced stripe widths.

In this experiment, the stripe width was varied $w \in \{30, 50, 70, 90, 110\}$ µm at constant resonator length $L = 4\,\text{mm}$ and vertical design (EDAS Gen 2 SQW, cf. section 3.2.2). All emitters have weak index guiding trenches ($\Delta n_{eff} = 6 \times 10^{-5}$, residual layer thickness $d_{res} = 320\,\text{nm}$) that are placed at a distance $a = 35\,\text{µm}$ from the injection stripe edge (cf. figure 3.26). The basic LIV characterization is shown in figure 4.1 and the LIV results are summarized in table 4.1.

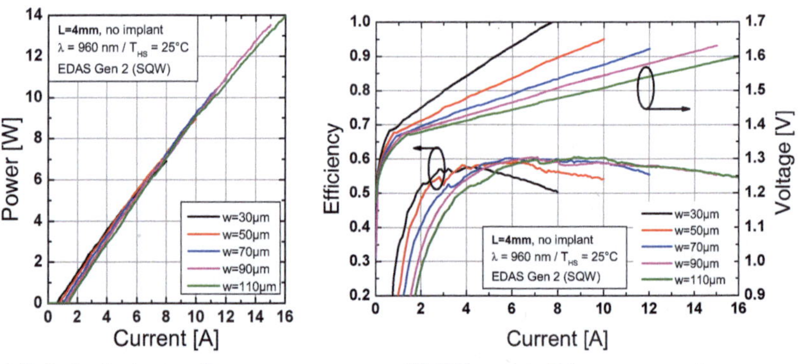

(a) Optical output power P_{opt} vs. current.

(b) Voltage and efficiency vs. current.

Figure 4.1: Basic LIV characterization of broad area emitters with varying stripe width. (a) Output power as function of injection current shows increasing threshold current I_{thr} for emitters with broader stripes. (b) Voltage and efficiency as function of injection current. With increasing stripe width, the series resistance decreases and the efficiency at high currents increases.

With increasing stripe width, the threshold current I_{thr} increases due to the increased active area. Simultaneously, the threshold current density j_{thr} decreases because the proportion of current that is laterally spread into regions without carrier clamping becomes more and more negligible, i.e. the relative current loss due to edge leakage is getting smaller. The series resistance R_s decreases due to the widened current path. The maximum efficiency $\eta_{c,max}$ is higher for broader stripes and peaks at 60.7% for $w = 90\,\mu m$ devices.

In general, the output power at peak efficiency $P_{\eta max}$ is higher for broader stripes and the efficiency function is less curved for higher currents. A quantification for this is the efficiency decrease rate at currents $I > I_{\eta max}$: $Z = \Delta\eta/\Delta I|_{\eta>\eta max}$. This value is highest for the relatively narrow stripe widths $w = 30\,\mu m$ and $w = 50\,\mu m$. The lowest Z is reached at $w = 70\,\mu m$ and $w = 90\,\mu m$, with a slight increase for $w = 110\,\mu m$. In summary, BALs with broader stripes reach higher output powers at higher efficiencies and the efficiency drop at high currents is slower as for BALs with narrow stripe widths.

parameter	unit	stripe width w				
		$30\,\mu m$	$50\,\mu m$	$70\,\mu m$	$90\,\mu m$	$110\,\mu m$
I_{thr}	[mA]	554	730	924	1183	1279
j_{thr}	[Acm^{-2}]	462	365	330	329	291
S	[W A^{-1}]	1.02	1.03	1.05	1.04	1.03
$\eta_{c,max}$	[%]	57.5	59.5	60.4	60.7	60.1
R_s	[mΩ]	44.7	30.1	22.7	18.9	15.9
R_{th}	[K W^{-1}]	4.7	3.7	2.75	2	2.42
U_0	[V]	1.361	1.355	1.354	1.354	1.351
$I_{\eta max}$	[A]	4.3	6.1	5.9	7.3	8.8
$P_{\eta max}$	[W]	3.9	5.6	5.2	6.6	7.8
Z	[A^{-1}]	-0.021	-0.013	-0.005	-0.005	-0.007

Table 4.1: Summarized results from LIV-measurement on BALs with varying stripe width w.

The lateral beam characteristics are depicted in figure 4.2. With increasing stripe width w, the near field width $w_{95\%}$ increases as well, but is generally narrower than w - especially at increased output power, as shown in figure 4.2(a). The stripe width dependent far field divergences in figure 4.2(b) describe a clear tendency: For smaller values of w, the far field angle is increased for equal power levels and grows faster with increasing power. Except for the BAL with $w = 30\,\mu m$, whose divergence growth is superlinear, all configurations show a linear growth $\Delta\Theta_{95\%}/\Delta P_{opt}$ of the far field divergence with output power. The values are noted in table 4.2 and for the broadest stripe width in test ($w = 110\,\mu m$) that growth rate decreases to $(0.63 \pm 0.03)\,^\circ W^{-1}$.

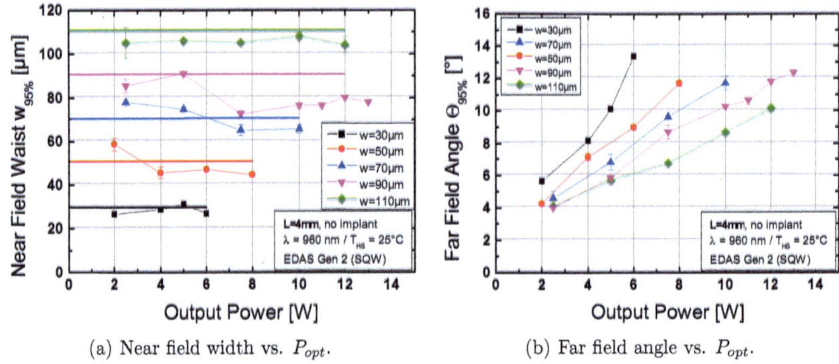

(a) Near field width vs. P_{opt}.　　　　　(b) Far field angle vs. P_{opt}.

Figure 4.2: Near- and far field dimensions as function of output power P_{opt} for BALs with varying stripe width. (a) The near field widths decrease with increasing output power due to thermal lensing. (b) With increasing stripe width, the slope of the far field broadening decreases.

The near- and far field widths are converted into BPP_{lat} values that are presented in figure 4.3. Even though the far field divergence is increased for narrow stripes, the overall BPP_{lat} is lower compared to broad stripes due to the reduced near field width. Interestingly, there is almost no difference between BALs with $w = 70\,\mu\text{m}$ and $w = 90\,\mu\text{m}$. However, for BALs with $w = 110\,\mu\text{m}$, BPP_{lat} grows again.

In figure 4.3(b), the lateral beam parameter product is plotted as a function of active zone temperature increase ΔT_{AZ}. With decreasing stripe width, the BPP_0 ground level and the thermal slope S_{th} decrease, as noted in table 4.2.

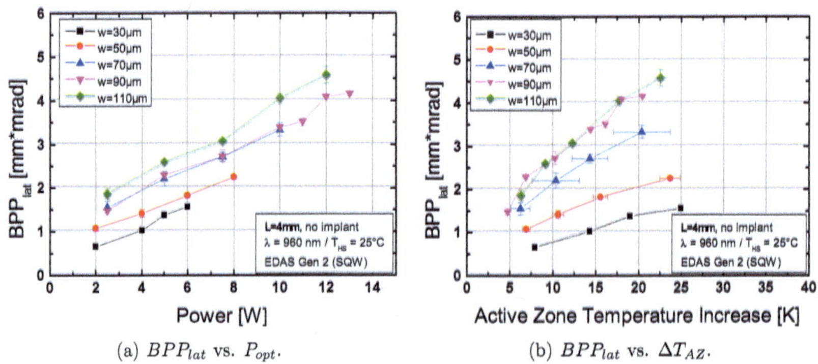

(a) BPP_{lat} vs. P_{opt}.　　　　　(b) BPP_{lat} vs. ΔT_{AZ}.

Figure 4.3: Lateral beam parameter product BPP_{lat} of BALs with varying stripe width w as function of optical output power P_{opt} and ΔT_{AZ}. (a) $BPP_{lat}(P_{opt})$ function shifts to higher BPP-values with increasing stripe width. (b) A reduced stripe width lowers the thermal slope S_{th} and BPP_0. For detailed comparison, see table 4.2.

94

parameter [unit]	stripe width w				
	$30\,\mu m$	$50\,\mu m$	$70\,\mu m$	$90\,\mu m$	$110\,\mu m$
$\Delta\Theta_{95\%}/\Delta P_{opt}$ [° W^{-1}]	1.9 ± 0.3	1.21 ± 0.05	0.99 ± 0.06	0.82 ± 0.02	0.63 ± 0.03
BPP_0 [mm mrad]	0.2 ± 0.1	0.7 ± 0.2	0.8 ± 0.1	0.9 ± 0.3	1.1 ± 0.1
S_{th} [mm mrad K^{-1}]	0.05 ± 0.01	0.07 ± 0.01	0.13 ± 0.01	0.17 ± 0.02	0.16 ± 0.01

Table 4.2: Summarized results from beam quality measurement of BALs with varying stripe width w.

The BPP_0 and S_{th} decrease is supposed to be a consequence of a reduced number of guided lateral modes in narrower stripes. Even though the above mentioned cutoff-condition $q = W/\pi$ is only valid for strongly index-guided BALs, a reduced number of lateral modes can be assumed for narrow stripes with quasi-index guiding, as recently shown for $w = 30\,\mu m$ DFB-BALs [95].

In order to substantiate this assumption, a simulation based investigation of the guided modes in a thermal waveguide was performed. Therefore, a series of five thermal lenses for the stripe widths $w \in \{30, 50, 70, 90, 110\}$ µm was computed using the FEM-solver ANSYS [96]. The simulation domain contains a 2D-model (lateral-vertical plane) of the laser chip (EDAS Gen 1), its CuW carrier and the solder material, as described in [97]. However, for the purpose of this study, only the lateral temperature profile along the active zone is used. In figure 4.4(a), the series of thermal profiles is plotted. For all stripe widths, the operating point is selected so that the internal temperature increase is kept at $\Delta T_{AZ} \approx 15\,K$.

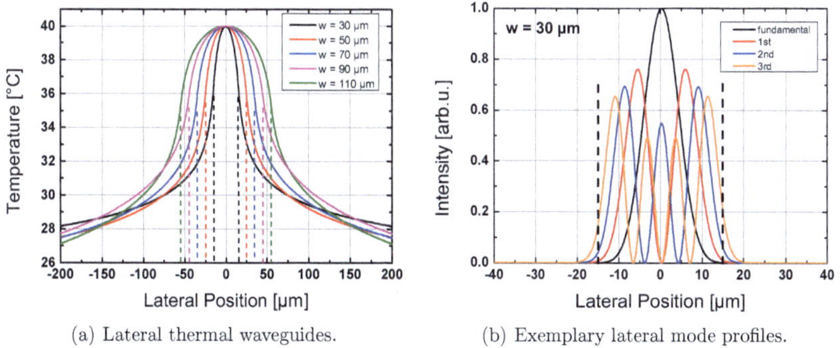

(a) Lateral thermal waveguides.

(b) Exemplary lateral mode profiles.

Figure 4.4: Lateral thermal lens, and electric field simulation in BALs with varying stripe width w. (a) Simulated lateral temperature profiles of BAL emitters with varying stripe width. The temperature increase is constant at $\Delta T_{AZ} = 15\,K$. (b) Calculated lateral modes (first four) guided by the $w = 30\,\mu m$ thermal waveguide. The dashed lines indicate the injection stripe width.

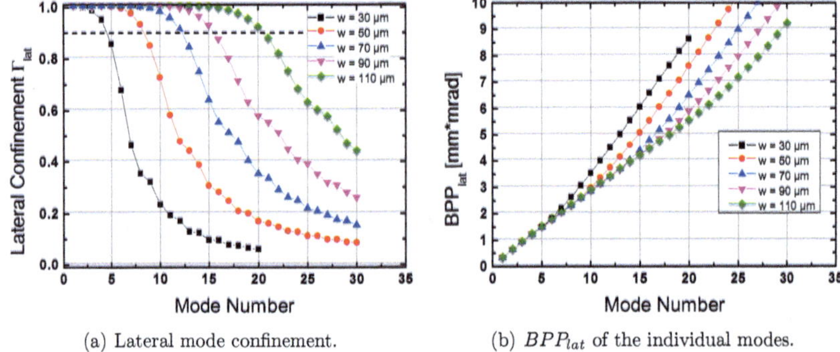

(a) Lateral mode confinement.   (b) BPP_{lat} of the individual modes.

Figure 4.5: Evaluation of the lateral mode properties in thermal waveguides of different width. (a) One-dimensional confinement Γ_{lat} of the lateral modes, assuming rectangular-shaped current injection into the active zone. (b) Lateral beam parameter product of each individual mode.

To obtain the guided modes, every temperature profile is converted into a refractive index profile using the model from Gehrsitz *et al.* [47]. This profile is used as the basis for the calculation of the lateral modes with a 1D Helmholtz solver, written in MatLab, that solves equation 2.15 (cf. section 2.2.2). As an example, the first four modes (near field intensity) of the $w = 30\,\mu m$ emitter are presented in figure 4.4(b).

For the evaluation, the mode overlap with the injection stripe and the mode beam parameter product is derived. Since modes with reduced overlap in the pumped region suffer from strong optical losses (free carrier absorption), their gain decreases. Finally, when the losses exceed the gain of the mode, it ceases to lase. Hence, the lateral mode confinement Γ_{lat} is directly correlated to the mode number in the sense that below a certain confinement threshold, no more modes start lasing. As depicted in figure 4.5(a), the lateral confinement decreases faster with increasing mode number for smaller stripe widths. As a consequence, it holds that for any given confinement threshold $\Gamma_{lat,thr} < 1$ the number of modes N_m that fulfill $\Gamma_{lat} > \Gamma_{lat,thr}$ is smaller for narrow injection stripes, such that $N_{m,w} < N_{m,\tilde{w}}$ for all stripe widths w and $\tilde{w}$ that fulfill the condition $w < \tilde{w}$.

The calculation of the lateral BPP is shown in figure 4.5(b). For every stripe width w, BPP_{lat} increases monotonically with the mode number. However, it does not hold that equal mode index results in equal BPP_{lat}. Beginning with mode number six, the BPP curves for different stripe widths start to diverge.

For the comparison with the measured beam properties, the following procedure is applied to each set of modes. First, a lateral confinement threshold is chosen. According to [42] that value is $\Gamma_{lat,thr} \approx 0.97$, however, here this threshold is set to $\Gamma_{lat,thr} = 0.9$ (dashed

96

line in figure 4.5(a)), since broader stripe widths are examined in this study and the index guiding trenches are relatively shallow, so that a broadened lateral gain region is assumed.

Second, all near- and far field intensity profiles for the modes that fulfill $\Gamma_{lat} > \Gamma_{lat,thr}$ are superposed uniformly, i.e. the contribution to the total intensity profile is equal for each mode. This assumption is, of course, arbitrary, since the exact distribution of power among the modes is not known. However, in the chosen range of Γ_{lat} almost all modes have 100% overlap with the gain region, so that a uniform distribution is reasonable. Then the widths with 95% power content are derived for all intensity profiles and the resulting BPP_{lat} is calculated (cf. table 4.3). In figure 4.6, the comparison of the beam profile simulation with the measured values is shown. For the near- and far field dimensions, the conformity with the measured data is good up to $w = 70\,\mu m$, with slight deviations in the far field angle prediction. For $w = 90\,\mu m$, the simulation estimates a $9\,\mu m$ broader near field and a 1° narrower far field (BPP_{lat} is predicted correctly, however). The far field divergence in the $w = 110\,\mu m$ BAL is predicted to be 2° larger than measured, which leads to a $1\,mm\,mrad$ increase in the predicted BPP_{lat}.

The far field deviations for broader injection stripes (cf. figure 4.6(a), profiles shown in figure 4.6(c)) are probably correlated to the assumption of uniform power distribution among all modes, which is expected to be increasingly inaccurate since the number of participating modes increases with w. This increase is quantified here (cf. table 4.3) as a mean of the quotients N_m/w and amounts to 0.16 modes per micron (basis: nominal injection stripe width). From figure 4.6(b), the predicted BPP_{lat} deterioration rate with stripe width is determined with a linear regression: $\Delta BPP_{lat}/\Delta w = (0.044 \pm 0.002)\,mm\,mrad/\mu m$, which is consistent with the measured BPP_{lat} increase up to $w = 90\,\mu m$.

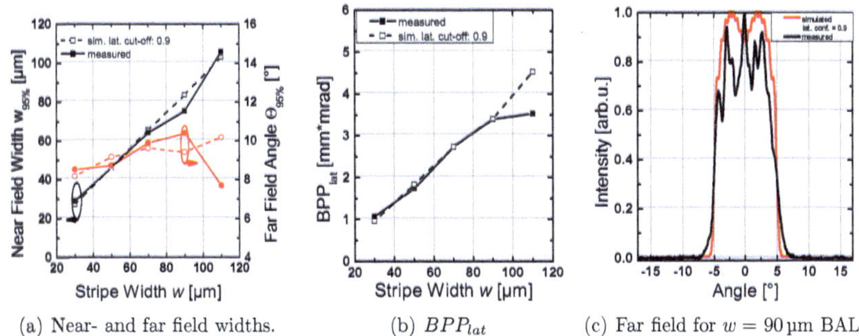

(a) Near- and far field widths.   (b) BPP_{lat} (c) Far field for $w = 90\,\mu m$ BAL

Figure 4.6: Comparison of predicted and measured intensity profile widths (a) and the resulting BPP_{lat} (b) as function of w. The calculation includes the modes that are guided by a $\Delta T_{AZ} = 15\,K$ thermal waveguide and have more than 90% lateral confinement. (c) The comparison of the far field profiles shows that the divergence is underestimated in $w = 90\,\mu m$ BALs

97

parameter	unit	stripe width w				
		30 µm	50 µm	70 µm	90 µm	110 µm
N_m	-	4	8	12	15	20
N_m/w	$[\mu m^{-1}]$	0.13	0.16	0.17	0.16	0.18
$w_{95\%,sim}$	$[\mu m]$	27	46	66	84	103
$w_{95\%}$	$[\mu m]$	29	46	64	75	106
$\Theta_{95\%,sim}$	$[°]$	8.1	9.1	9.5	9.3	10.1
$\Theta_{95\%}$	$[°]$	8.4	8.6	9.8	10.3	7.6
$BPP_{lat,sim}$	$[mm\,mrad]$	1	1.8	2.7	3.4	4.5
BPP_{lat}	$[mm\,mrad]$	1.1	1.7	2.7	3.4	3.5

Table 4.3: Comparison of simulated and measured beam profiles at $\Delta T_{AZ} = 15\,K$ for varying stripe widths w. The simulated beam profiles include all modes that fulfill $\Gamma_{lat} > 0.9$ and uniform distribution of power among modes is assumed.

The results presented here indicate that the decrease in the lateral mode number is the cause of the decrease in BPP_0 and S_{th} with decreasing stripe width (cf. figure 4.3(b)). The simulation of lateral modes in a $\Delta T_{AZ} = 15\,K$ thermal waveguide in figure 4.5 clearly shows that BALs with narrow stripe widths guide fewer modes for the same junction temperature increase than BALs with broader stripe widths, so that the BPP_{lat} *must* increase in broader stripes due to the depicted BPP_{lat} growth with increasing mode number.

Moreover, the comparison presented in figure 4.6 showed that for increased stripe widths $w > 70\,\mu m$, the field distributions and the BPP_{lat} cannot be explained by simple uniform mode overlap.

Another possible cause for an S_{th} change could be a different level of LCA in BALs with different stripe widths. The LCA-effect is driven by current, but the diagram in figure 4.3(b) plots BPP_{lat} as a function of ΔT_{AZ}. Thus, it is worthwhile to have a look at how ΔT_{AZ} depends on current I and stripe width w.

For increased w, the series resistance R_s and the thermal resistance R_{th} are reduced (cf. table 4.1), so a higher current is needed to reach equivalent ΔT_{AZ} values. Is this current increase significant enough to cause a S_{th} increase via increased LCA? In order to assess this effect, an edge-region fill factor is introduced $FF_{edge} = (0.2 \cdot w[\mu m])/(w[\mu m] + 20)$, so that $I_{edge} = FF_{edge} \cdot I$. Here, based on the investigation in section 3.4, the total pumped area is of the width $w + 20$, accounting two zones of $10\,\mu m$ width beyond each stripe edge. According to the observation in figure 3.33(a), the pumped LCA-region is set to be 20 % of the stripe width. In equation 4.1, the dependence of ΔT_{AZ} on current I is shown:

$$\Delta T_{AZ}(I) = (U \cdot I - P_{opt}) \cdot R_{th} \cong ((U_0 \cdot I + R_s \cdot I^2) - (S \cdot (I - I_{thr}))) \cdot R_{th}. \qquad (4.1)$$

Except for the slope $S = 1.03\,\mathrm{W\,A^{-1}}$, which is fixed for all stripe widths, all parameters in equation 4.1 have a dependence on the stripe width that is given by fitting the data from the LIV characterization, as shown in figure 4.7.

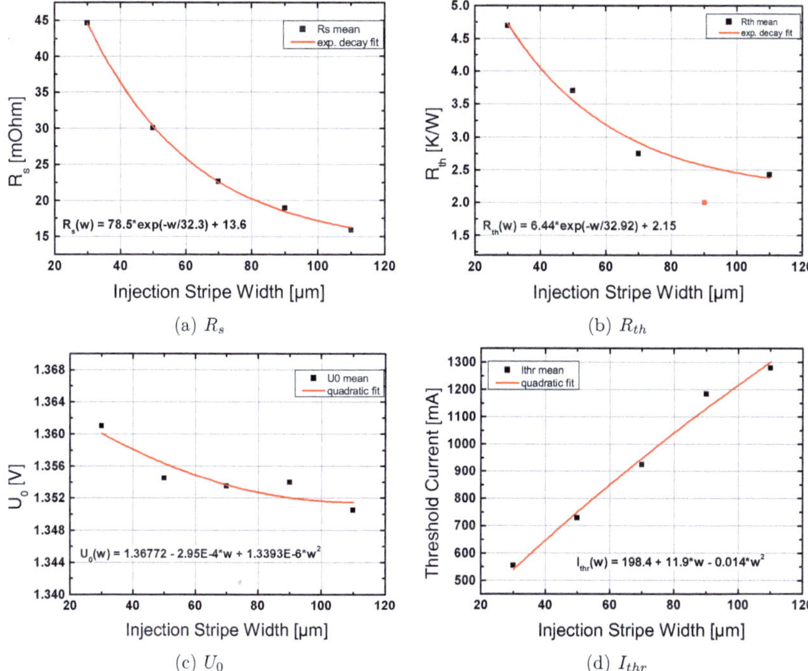

(a) R_s

(b) R_{th}

(c) U_0

(d) I_{thr}

Figure 4.7: Fitting curves for series resistance R_s (a), thermal resistance R_{th} (b), band gap voltage U_0 (c) and threshold current I_{thr} (d) as function of injection stripe width w. The fitting functions are chosen to yield the smallest deviation from the data and are not based on equations that intrinsically describe the behavior of these parameters. Please note: The R_{th} value for $w = 90\,\mu\mathrm{m}$ BALs is surprisingly low due to irregular wavelength drift. Therefore, it was excluded from the fitting procedure.

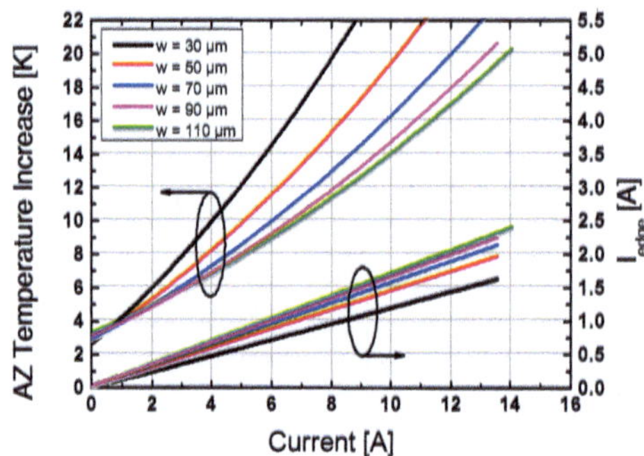

Figure 4.8: Stripe width dependent function of ΔT_{AZ}^{w} with current (left scale) and current proportion that reaches critical LCA-regions I_{edge} (right scale).

The fit-equations read:

$$R_s(w)[\text{m}\Omega] = 78.5 \cdot \exp(-w[\mu\text{m}]/32.3) + 13.6 \tag{4.2}$$

$$R_{th}(w)[\text{K W}^{-1}] = 6.44 \cdot \exp(-w[\mu\text{m}]/32.92) + 2.15 \tag{4.3}$$

$$U_0(w)[\text{V}] = 1.36772 - 2.95\text{E-4} \cdot w[\mu\text{m}] + 1.3393\text{E-6} \cdot w^2[\mu\text{m}^2] \tag{4.4}$$

$$I_{thr}(w)[\text{mA}] = 198.4 + 11.9 \cdot w[\mu\text{m}] - 0.014 \cdot w^2[\mu\text{m}^2] \tag{4.5}$$

Substituting parameter equations 4.2 - 4.5 into equation 4.1 yields the active zone temperature increase ΔT_{AZ}^{w} as a function of current and stripe width. In figure 4.8, this function is plotted for $w \in \{30, 50, 70, 90, 110\}$ µm. In the same diagram, the current that reaches the stripe-edge regions I_{edge} is plotted.

Now, for every temperature increase ΔT_{AZ}, the driving currents I and I_{edge} can easily be inferred. The list of edge currents in table 4.4 shows that with increasing stripe width w, the current proportion that reaches the critical region for LCA increases as well.

Comparing the smallest stripes ($w = 30\,\mu$m) with the broadest stripes ($w = 110\,\mu$m), this increase is more than 100 % for every ΔT_{AZ}.

This implies that a part of the S_{th} increase with increasing w is caused by stronger current injection into the stripe edges. However, it should be noted here that this result is strongly dependent on the choice of FF_{edge}. If the numerator is fixed and independent of w, the result would be a *reduction* in I_{edge} with increasing w. But it seems intuitively right to assume a growing LCA-region with growing stripe width. The LCA effect is a possible explanation for the far field deviations observed for $w = 90\,\mu$m BALs in figure 4.6.

ΔT_{AZ} [K]	edge current I_{edge} [A]				
	$30\,\mu$m	$50\,\mu$m	$70\,\mu$m	$90\,\mu$m	$110\,\mu$m
5	0.18	0.28	0.35	0.38	0.39
10	0.48	0.74	0.96	1.09	1.18
15	0.74	1.13	1.46	1.67	1.81
20	0.98	1.48	1.9	2.16	2.35

Table 4.4: Current that reaches edge regions I_{edge} for different active zone temperatures and varying stripe width w, assuming an edge fill-factor of $FF_{edge} = 0.2w/(w+20)$.

Conclusions

The present study showed that BALs with wider injection stripes have a higher efficiency at increased operation currents, an increased maximum output power and increased BPP_{lat}. The widened current aperture reduces the thermal- and series resistance. However, as the thermal waveguide is widened as well, the number of guided lateral modes N_m increases at a rate of $N_m/w \approx 0.16$ modes per micron and causes deterioration of the beam quality.

Based on thermal waveguide and lateral mode simulations, the BPP_{lat} deterioration rate was estimated to be $\Delta BPP_{lat}/\Delta w = (0.044 \pm 0.002)\,$mm mrad$/\mu$m (valid for 15 K active zone temperature increase). In addition there are indications that the beam quality is further decreased by an increased level of LCA, especially for BALs with $w > 90\,\mu$m.

The growth of BPP_{lat} for broader stripe widths exceeds the growth in output power. Hence, the linear radiance $B_{lin} = P_{opt}/BPP_{lat}$ decreases with increasing w, as shown in figure 4.9.

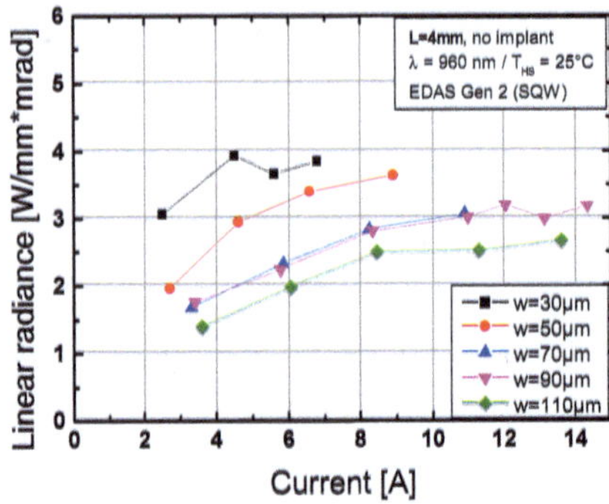

Figure 4.9: Linear radiance B_{lin} for BALs with varying stripe width w. Broader stripes show higher output power, but high BPP_{lat} brings B_{lin} down. Highest radiance is achieved for narrow stripes with $w = 30\,\mu$m.

In this investigation, the maximum radiance is reached for $w = 30\,\mu$m BALs at $P_{opt} = 4\,$W and $B_{lin} = 4\,$W/mm mrad.

Now, in order to obtain BAL emitters with high radiance, the following strategy is deduced. Since the output power at maximum efficiency is limited in narrow injection stripes, the strategy is to apply mode filtering techniques that suppress higher order lateral modes in BALs with *broad* injection stripes in order to reduce BPP_{lat} and maintain high optical output. The following section presents one approach that is based on the observation that higher order lateral modes have a larger near field expansion $w_{95\%}$.

4.2 Anti-Waveguide Layers as lateral mode filter

As illustrated in figure 4.5(a), an increase in mode number is inevitable when the stripe width of broad area lasers is increased. In order to avoid a drastic BPP_{lat} increase, while maintaining high efficiency and output power, different mode-filtering techniques are applied in the BAL design. Here, the goal is to tailor the mode losses in a way that higher order lateral modes suffer strong losses and do not reach threshold.

In ridge waveguide lasers, mode filtering in lateral direction has been applied in order to increase the kink-free output power and allow fundamental mode emission at higher bias. Examples of filtering techniques include the creation of highly resistive layers via hydrogen passivation of the regions beneath the ridge [98] or ridge-corrugation [99]. In broad area lasers, similar techniques are possible. However, the 'boundary-condition' is that neither efficiency nor reliability are compromised.

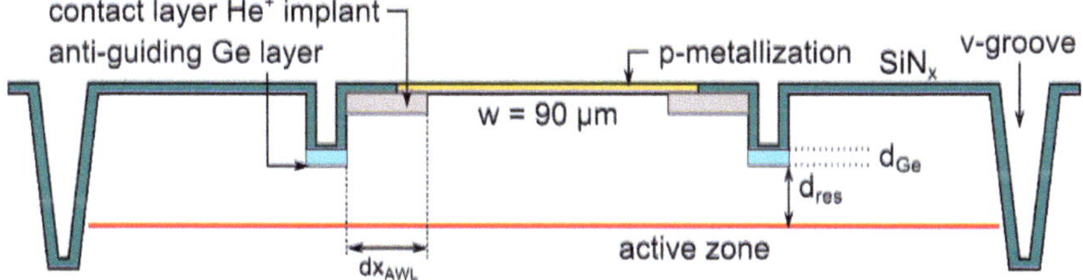

Figure 4.10: Design of BALs with 'anti-waveguide' layers in lateral-vertical cross-section. Important design parameters for performance optimization are the residual layer thickness d_{res}, the germanium layer thickness d_{Ge} and the anti-waveguide distance dx_{AWL}.

The mode filtering approach that will be presented here was proposed by Wenzel *et al.* [100]. It relies on the resonance between two congruent vertical waveguides, one of which contains a layer that causes strong optical absorption. The adaptation of the effective refractive index n_{eff} in both waveguides allows resonant out-coupling of the electric field and the strength of the coupling can be regulated by a geometrical offset.

The lateral BAL design used in this study is presented in figure 4.10. Here, the central vertical waveguide in the stripe region is surrounded by two 'anti-waveguides'. The latter are fabricated as dry etched trenches refilled with germanium, which has a strong absorption of $\alpha_{Ge} \approx 39\,000\,\text{cm}^{-1}$. Without the germanium, the effective index step from center to anti-waveguide is determined by the residual layer thickness d_{res}. An increased d_{res} gives a lower effective index difference Δn_{eff}, which results in a reduced number of guided modes. At the same time, however, the calculated losses for higher order modes becomes weaker with increased d_{res} [100].

The thickness of the germanium layer d_{Ge} is used to adjust the effective index of the fundamental mode in the anti-waveguide region to match that of the central waveguide. As a diagnostic tool, the distance between the anti-guiding layer and the stripe edge is varied $dx_{AWL} \in \{5, 10, 15\}$ μm.

The functional principle of the BALs with anti-waveguide can be described as follows. As soon as the current is increased beyond threshold, lasing begins with a few lower-order modes which are well confined in the central waveguide. With further increased current, the thermal lens builds up and an increasing number of higher order modes start lasing. But the higher the mode number, the wider its near field becomes (cf. figure 4.4(b)). This expansion towards the stripe edge increases the penetration of the higher order modes into the anti-waveguide, which is optically not separated from the central waveguide since it was designed to have the same effective index n_{eff} (i.e. $\Delta n_{eff} = 0$). The greater the intensity that is coupled into the anti-waveguide, the higher the loss for that individual mode due to absorption in the Ge-layer. In this way, the lasing of higher order modes with high BPP_{lat} (cf. figure 4.5(b)) is suppressed and the overall beam quality is improved.

In general, the AWL filters modes within an n_{eff} range, but since in BALs all modes have a similar n_{eff}, the filtering relies on the near field expansion. However, if all lateral modes show a comparable expansion, e.g. in structures with lateral index guiding (cf. figure 3.27(a)), then all modes suffer from strong absorption losses and the efficiency will be compromised.

Experimental results

AWL in ASLOC structure

In a first proof of principle test, the anti-waveguide design was applied to an SQW-ASLOC structure comparable to the vertical design presented in section 3.2. The total p-side thickness equals $d_{p-side} = 1965\,$nm, containing a 950 nm thick p-waveguide and a 500 nm p-cladding. The trenches were dry etched at a distance of $dx_{AWL} = 10\,\mu$m with a residual layer thickness $d_{res} = 1100\,$nm, which results in an effective index step of $\Delta n_{eff} = 1E{-}5$. The trenches were filled with germanium and the calculated resonance was at $d_{Ge} = 43\,$nm. The test devices were processed as BAL-emitters with a resonator length $L = 4\,$mm and a stripe width $w = 90\,\mu$m, emitting at $\lambda = 973\,$nm.

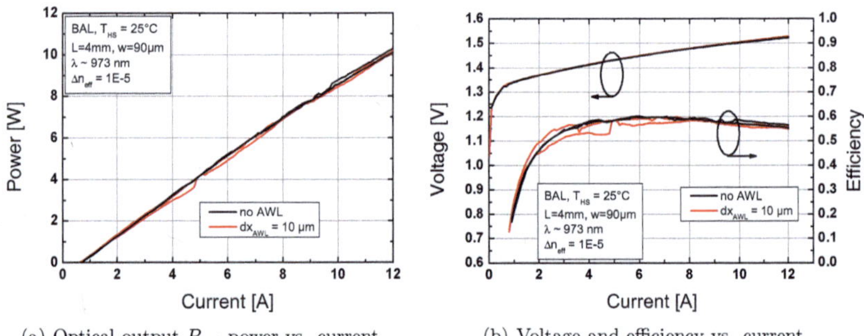

(a) Optical output P_{opt} power vs. current.   (b) Voltage and efficiency vs. current.

Figure 4.11: Basic LIV characterization of ASLOC BAL emitters with and without anti-waveguide layers. All emitters have resonator length $L = 4\,$mm and stripe width $w = 90\,\mu$m. The AWL is placed at $dx_{AWL} = 10\,\mu$m. (a) Both designs show equivalent maximal output power of $P_{max} = 10\,$W at $I_{max} = 12\,$A. (b) The efficiency is slightly reduced in AWL BALs, especially for $I < 8\,$A.

In figure 4.11, the impact of the anti-waveguide on the output power and conversion efficiency is shown. It should be noted, that the BALs with AWL show more kinks in their power curve - an effect that was previously observed in [100]. The efficiency curve shows corresponding irregularities for AWL-BALs. For $I < 8\,\mathrm{A}$, the efficiency is reduced (decrease $\leq 5\%$ in one emitter) in devices with antiwaveguide. However, both the reference without germanium and the AWL-BAL have the same maximum output power at $P_{max} = 12\,\mathrm{W}$.

The results of the beam quality assessment are presented in figures 4.12 and 4.13. A considerable reduction in the near field waist is observed for AWL BALs. This reduction extends over a large power-range from $2\,\mathrm{W}$ to $10\,\mathrm{W}$ and amounts to $\approx 20\%$. The far field angle with 95% power content is also decreased in the anti-waveguide design. For output powers $P_{opt} < 10\,\mathrm{W}$ the decrease ranges between $1°$ and $2.4°$, depending on the power level. At $P_{max} = 10\,\mathrm{W}$, however, the far field divergence for both structures reaches $\Theta_{95\%} = 10°$.

To further assess the difference between the near- and far fields for emitters with AWL, a representative profile comparison for $P_{opt} = 6\,\mathrm{W}$ is shown in figure 4.13. As shown by the near field comparison in figure 4.13(a), the decrease in profile width stems from a suppression of side-lobe intensity. The reference BAL near field has two pronounced side lobes at $\pm\,57\,\mu\mathrm{m}$, which vanish in the AWL-BAL near field. Furthermore, the far field profile of the AWL-BAL shows less pronounced shoulders at $\pm\,5°$. This observation is in good agreement with the working principle of the anti-waveguide. Reduced near field side-lobes and far field shoulders indicate the suppression of higher-order lateral modes, since these have a broad lateral extension.

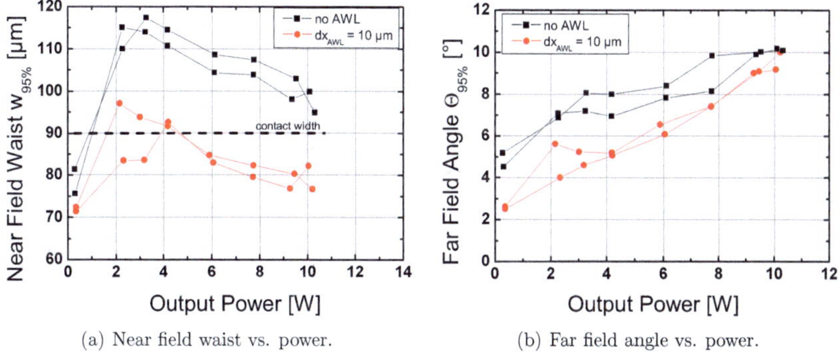

(a) Near field waist vs. power.

(b) Far field angle vs. power.

Figure 4.12: Optical field dimensions for ASLOC AWL- and reference BALs. (a) The near field width of AWL-BALs is reduced by $\approx 20\,\mu\mathrm{m}$. (b) The far field divergence of AWL-BALs is reduced by $\approx 1.7°$ for powers $P_{opt} < 10\,\mathrm{W}$.

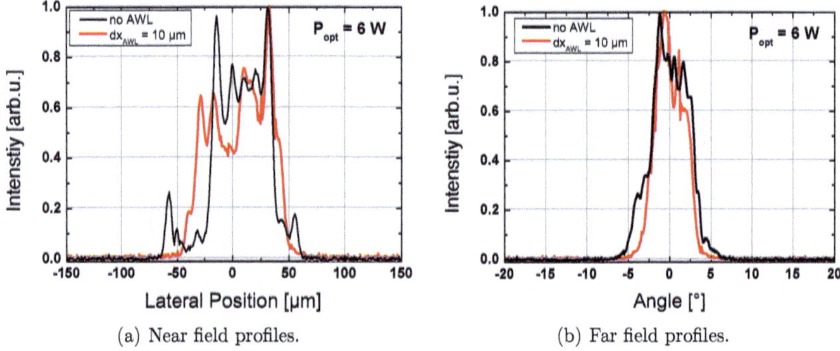

(a) Near field profiles.   (b) Far field profiles.

Figure 4.13: Near- and far field profiles of ASLOC AWL- and reference BALs. (a) The near field reduction in AWL-BALs stems from a suppression of side lobes. (b) A reduction in far field divergence is seen in AWL-BALs. The shoulders of the far field profile are less pronounced and steeper.

The decrease in near- and far field width for AWL-BALs also leads to a decrease in BPP_{lat}, as shown in figure 4.14. Over the power range of 4 W - 10 W, the BPP is reduced by 1.5 mm mrad - 1 mm mrad. The evaluation of the BPP_{lat} increase over ΔT_{AZ} yields a drop in BPP_0 from (3.3 ± 0.3) mm mrad to (1.3 ± 0.1) mm mrad (-60%) and a simultaneous increase in S_{th} from (0.06 ± 0.02) mm mrad K^{-1} to (0.10 ± 0.01) mm mrad K^{-1} (+66%).

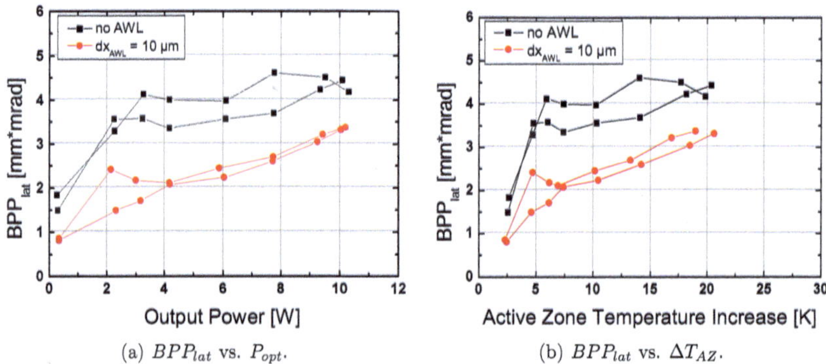

(a) BPP_{lat} vs. P_{opt}.   (b) BPP_{lat} vs. ΔT_{AZ}.

Figure 4.14: Lateral beam parameter product of ASLOC BALs as function of optical output power (a) and active zone temperature increase (b). The reduced near field width results in reduced BPP_{lat} over the whole power-range. The BPP ground level BPP_0 decreases by 2 mm mrad in BALs with antiwaveguides at $dx_{AWL} = 10 \,\mu$m.

In summary, the proof of principle test was successful. The application of two anti-waveguides in a SQW-ASLOC alongside the injection stripe at $dx_{AWL} = 10 \,\mu$m led to a considerable reduction in the BAL BPP_{lat} of up to 1.5 mm mrad, by eliminating the near field side lobes and reducing far field profile shoulders. However, in these emitters the near field widths are very large to begin with, since the non-AWL structure has a near field that exceeds the contact width of $90\,\mu$m (dashed line in figure 4.12(a)) by $\approx 20\,\mu$m. Hence, the ASLOC AWL-BALs show improved the linear radiance, but do not exceed previously presented BAL results (cf. section 3.4.2), with $B_{lin} = 3\,\mathrm{W/mm\,mrad}$.

AWL in EDASLOC structure

In order to check the compatibility of the anti-waveguide design with other vertical structures, a further diagnostic test with EDAS Gen 2 (DQW) emitters was performed. As a free parameter, the AWL distance was varied $dx_{AWL} \in \{5, 10, 15\}$ µm. As in the first test, all emitters have resonator length $L = 4\,\mathrm{mm}$ and stripe width $w = 90\,\mu$m.

In EDAS Gen 2, the p-side is $d_{p-side} = 1980\,\mathrm{nm}$ thick. A thickened sub-contact layer (800 nm) is followed by a p-cladding that consists of $800\,\mathrm{nm}$ $\mathrm{Al_{0.85}Ga_{0.15}As_{0.98}P_{0.02}}$ and the p-waveguide, which is a 220 nm thick GRIN-layer. The trenches were etched to yield an effective index difference of $\Delta n_{eff} = 1.1 \times 10^{-5}$, as in the previous test. To reach this index step, the residual layer thickness had to be reduced to $d_{res} = 420\,\mathrm{nm}$, due to the strong asymmetry of p- and n-side in the EDAS structure. The trench offset is fixed at $a = 8\,\mu$m, except for the emitters with $dx_{AWL} = 5\,\mu$m (here $a = 5\,\mu$m).

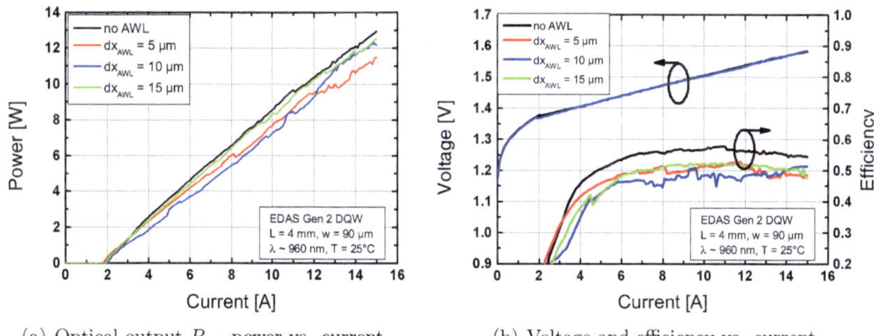

(a) Optical output P_{opt} power vs. current.   (b) Voltage and efficiency vs. current.

Figure 4.15: Basic LIV characterization of EDASLOC BAL emitters with anti-waveguide layers at different lateral distances $dx_{AWL} \in \{5, 10, 15\}$ µm. Results from representative emitters are shown. (a) Slope and maximal output power P_{max} are reduced in AWL-BALs and differ with dx_{AWL}. Furthermore, several kinks are observed as soon as the anti-waveguide is applied. (b) A considerable efficiency-penalty of at least 5% is observed in the AWL BALs . The efficiency decrease does not scale with dx_{AWL}.

In figure 4.15, the power, voltage and efficiency curves are shown as function of current. The power curve in figure 4.15(a) shows that the devices with anti-waveguide have a reduced slope, dropping from $1.03\,\mathrm{W\,A^{-1}}$ (reference) to $0.91\,\mathrm{W\,A^{-1}}$ for the two closest AWL configurations at $dx_{AWL} = 5\,\mu\mathrm{m}$ and $10\,\mu\mathrm{m}$. For the remote configuration at $dx_{AWL} = 15\,\mu\mathrm{m}$ the slope recovers to $1.02\,\mathrm{W\,A^{-1}}$ (all values are based on a mean of 3 emitters per species).

The maximal power P_{max} follows a similar trend. The reduction in P_{max} is $1.2\,\mathrm{W}$ for the minimal AWL distance at $dx_{AWL} = 5\,\mu\mathrm{m}$. At $dx_{AWL} = 10\,\mu\mathrm{m}$ and $dx_{AWL} = 15\,\mu\mathrm{m}$, the power penalty decreases to $\approx 0.4\,\mathrm{W}$. So, regarding P_{max}, the performance difference between regular and AWL-BAL decreases with increasing dx_{AWL}. However, the efficiency curve (cf. figure 4.15(b)) does not confirm this tendency, since here the AWL BALs with $dx_{AWL} = 10\,\mu\mathrm{m}$ have the lowest conversion efficiency ($\Delta\eta_c = 3\text{-}11\%$) over a large current range $I < 12\,\mathrm{A}$. A further observation is that - again - all power and efficiency curves for BALs with the anti-waveguide show considerable fluctuations (kinks).

Regarding the near field dimensions as function of output power in figure 4.16(a), the behavior is different to the previously assessed ASLOC BALs. Here, the anti-waveguide layers do not reduce the near field width. Instead, no clear trend is separable due to the chip-to-chip scattering (indicated by the vertical error bars), which is very distinct.

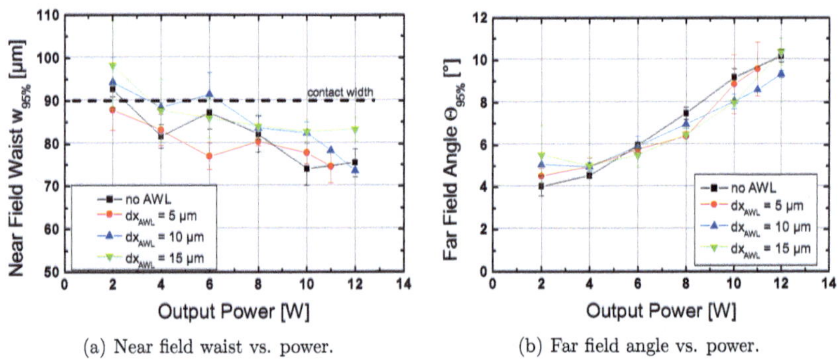

(a) Near field waist vs. power.

(b) Far field angle vs. power.

Figure 4.16: Near- and far field dimensions of EDASLOC AWL BALs as function of output power for different anti-waveguide distances dx_{AWL}. (a) For all emitters, the typical near field shrinkage at higher output powers is observed. However, no clear separation between regular and AWL BALs is seen. The near field width is not correlated to dx_{AWL}. (b) The AWL BALs with $dx_{AWL} = 10\,\mu\mathrm{m}$ and $15\,\mu\mathrm{m}$ show a reduced far field divergence by $\approx 1°$ for $P_{opt} > 6\,\mathrm{W}$.

In the corresponding plot for the far field divergence (c.f. figure 4.16(b)), a similar observation is made. For $P_{opt} \leq 6\,\mathrm{W}$, no difference between the regular BAL and the AWL BAL can be distinguished. However, for $P_{opt} > 6\,\mathrm{W}$, the emitters with $dx_{AWL} = 10\,\mu\mathrm{m}$

show a constant reduction in the far field divergence. Emitters with $dx_{AWL} = 15\,\mu m$ show a similar trend, but here the reduction vanishes at $P_{opt} = 12\,W$.

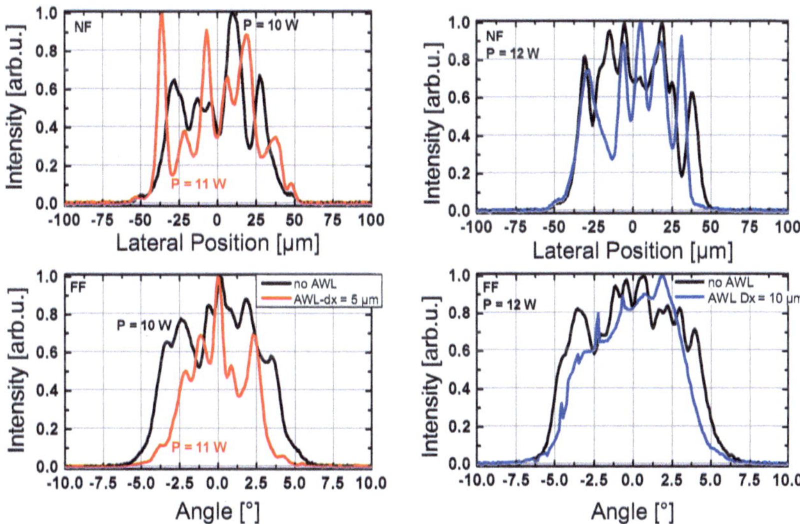

(a) Near and far field profiles for $dx_{AWL} = 5\,\mu m$. (b) Near and far field profiles for $dx_{AWL} = 10\,\mu m$.

Figure 4.17: Intensity profiles for near fields (upper row) and far fields (lower row) for AWL BALs with $dx_{AWL} = 5\,\mu m$ (red, left column) and $dx_{AWL} = 10\,\mu m$ (blue, right column) in comparison with reference profiles (black). Both AWL BAL emitters show a stronger near field modulation than the reference and reduced far field divergence.

A closer look at the intensity profiles in figure 4.17 shows that the near field profiles for BALs with anti-waveguide have a stronger modulation (near field contrast) than the reference. The AWL BAL with $dx_{AWL} = 5\,\mu m$ has $c_{NF} = 2.5$, which is a 47%-increase compared to the reference with $c_{NF} = 1.7$. For the emitter with $dx_{AWL} = 10\,\mu m$, the difference is smaller. Here a 32%-increase is observed from the reference value $c_{NF} = 0.7$ to $c_{NF} = 0.93$ for the AWL BAL. A reduction in the near field width, as shown for the emitter with $dx_{AWL} = 10\,\mu m$ in figure 4.17(b), is rather exceptional. In comparison to the ASLOC emitters, the non-AWL near fields of the EDAS-emitters show no pronounced side lobes, which is due to the reduced residual layer thickness d_{res}, that was reduced by 62% down to 420 nm. This brings the AWL closer to the central stripe region, increasing the out-coupling strength. In addition, due to the d_{res} reduction the lateral carrier spreading is reduced and the gain for laterally extended higher order modes decreases. As a consequence, the BAL near field contains fewer modes and is more compact, so that the AWL also extracts energy from low order modes and compromises the conversion efficiency, as seen in figure 4.15(a).

However, as shown in figure 4.17, the far field divergence is reduced as well, which is the most pronounced effect of the AWL in this study with EDAS material.

The near- and far field widths were converted into BPP_{lat} values, which are shown as function of power in figure 4.18(a). Unlike the test with ASLOC BALs, no clear distinction between reference emitters and AWL BALs is seen. All curves lie closely together and chip-to-chip scattering makes it difficult to derive useful conclusions. At $P_{opt} = 6\,\text{W}$ and $8\,\text{W}$, the AWL at $dx_{AWL} = 5\,\mu\text{m}$ and $15\,\mu\text{m}$ show a reduced BPP_{lat} by $\approx 0.5\,\text{mm\,mrad}$. At the highest power available ($P_{opt} = 12\,\text{W}$), the emitters with $dx_{AWL} = 10\,\mu\text{m}$ show the lowest BPP_{lat} at $3\,\text{mm\,mrad}$, which is a 12% reduction compared to the reference with $3.4\,\text{mm\,mrad}$. In contrast to this, the emitters with $dx_{AWL} = 15\,\mu\text{m}$ have an increased beam parameter product of $3.7\,\text{mm\,mrad}$.

In figure 4.18(b), the linear radiance B_{lin} of selected emitters is plotted as function of current. It shows that the anti-waveguide layers have the potential to improve the radiance significantly. At $I = 10\,\text{A}$, the emitters with $dx_{AWL} = 5\,\mu\text{m}$ and $10\,\mu\text{m}$ lift the radiance by $\approx 1\,\text{W/mm\,mrad}$. The 'hero'-emitter has $dx_{AWL} = 5\,\mu\text{m}$ and reaches $4.4\,\text{W/mm\,mrad}$ at $I = 14.5\,\text{A}$ and $\eta_c = 45.3\%$ conversion efficiency. However, no reliable statement on performance improvements can be given due to strong differences between laser chips which are nominally identical.

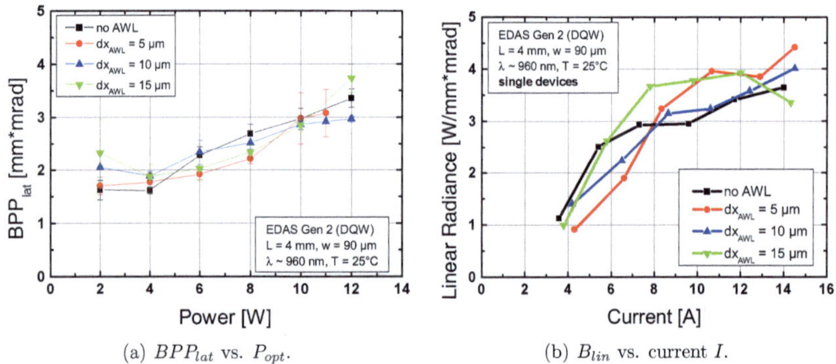

(a) BPP_{lat} vs. P_{opt}.

(b) B_{lin} vs. current I.

Figure 4.18: Beam quality and linear radiance of anti-waveguide BALs with varying dx_{AWL}. (a) The lateral beam parameter product as function of output power shows that the anti-waveguide layers improve BPP_{lat} for $4\,\text{W} < P_{opt} < 10\,\text{W}$. At $P_{opt} = 12\,\text{W}$, only the emitters with $dx_{AWL} = 10\,\mu\text{m}$ show a reliable improvement. (b) Linear radiance as function of current for selected emitters from every assessed configuration. The anti-waveguide can improve the maximal radiance, as shown by the emitters with $dx_{AWL} = 5\,\mu\text{m}$ (best radiance at $B_{lin} = 4.4\,\text{W/mm\,mrad}$) and $dx_{AWL} = 10\,\mu\text{m}$.

Conclusions

The study here showed that the anti-waveguide lateral mode filter can improve the beam quality of broad area emitters. The proof of principle test in an ASLOC vertical structure yielded a reduction in BPP_{lat} by 1.5 - 1 mm mrad and the conversion efficiency was reduced by at most 5%-points at currents $I < 8\,\mathrm{A}$.

However, a strong dependence on the vertical design in use was also revealed. The test of the anti-waveguide layer in an asymmetric vertical structure (EDAS) showed a significant efficiency loss (at least 3%-points at maximum current, $> 10\%$-points at lower bias), considerable chip-to-chip data spreading and increased near field contrast. The latter two observations indicate an instability of the design, making the chips prone to performance changes due to slight deviations in the design parameters. Here, the residual layer thickness d_{res} was found to be critical. It regulates the coupling strength and the near field width $w_{95\%}$, that determines how the AWL affects the lateral mode spectrum. In the above mentioned EDASLOC design d_{res} was reduced to 38% of its ASLOC value due to the thinned p-side. That reduction shifts the germanium layer closer to the active area and hence increases the absorption of the lateral modes, yielding a higher efficiency penalty and more irregularities in the power and efficiency curves.

In general, no clear dependence of the BAL efficiency, near - and far field width on dx_{AWL} was found. The benefit in BPP_{lat} has been achieved only for selected operating points, e.g. a reduction from 3.4 mm mrad to 3 mm mrad at $P_{opt} = 12\,\mathrm{W}$ for devices with $dx_{AWL} = 10\,\mu\mathrm{m}$. However, the highest radiance was measured in a $dx_{AWL} = 5\,\mu\mathrm{m}$ AWL BAL and amounts to 4.4 W/mm mrad.

The origin of the BPP_{lat} improvement is also dependent on the vertical design. In the assessed ASLOC BALs, the reduction was caused by a constant near field width reduction and a decrease in far field divergence for low output powers. However, in EDAS BALs the near field is not changed, but the far field divergence was reduced for high output powers.

Chapter 5

Summary and Outlook

The investigation of slow axis beam quality degradation in broad area lasers in this doctoral thesis will be summarized in the following paragraphs. In order to enable a quantification of *'beam quality degradation'*, an empirical model was used to describe the growth of BPP_{lat} with the internal (local) temperature increase ΔT_{AZ}:

$$\boldsymbol{BPP_{lat}(\Delta T_{AZ}) = BPP_0 + S_{th} \cdot \Delta T_{AZ}}.$$

In most cases, this linear dependence was an accurate description of BPP_{lat} deterioration. Here, a diagram of the identified factors influencing the BPP_0 ground level and the thermal slope S_{th} will be provided. Finally, an outlook on future activities that aim for BAL beam quality improvement will be given.

5.1   Results of root cause analysis

In chapter 3, the possible influences on BPP_{lat} were initially listed in an Ishikawa diagram (cf. figure 3.1). Next, several experiments, each consisting of a series of BALs, were conducted to identify the impact of each potential factor. Here, the BAL stripe width $w = 90\,\mu\mathrm{m}$ and resonator length $L = 4\,\mathrm{mm}$ were fixed. In figure 5.1 the findings of chapter 3 are summarized in an updated Ishikawa diagram.

It was shown that the **epitaxial structure** and the **lateral thermal lens profile** are strongly correlated. A comparison of two vertical designs with different asymmetry (EDAS vs. ASLOC, cf. section 3.2) brought the following insights. Thermographic measurements indicate that the thermal lens bowing (parameterized via a quadratic fit coefficient B_2) is directly connected to the thermal slope S_{th}, since the latter increases as $|\Delta B_2/\Delta T_{AZ}|$ increases. In this study, S_{th} increases by 100% in EDAS BALs compared to ASLOC BALs, while $|\Delta B_2/\Delta T_{AZ}|$ grows by 40%. A potential origin for this difference

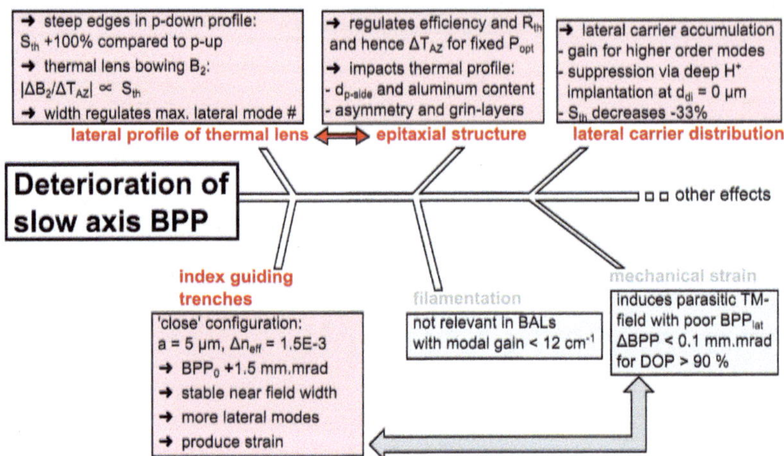

Figure 5.1: Initial list of factors that potentially influence the BPP_{lat} deterioration and findings of the root cause analysis. Important factors are marked in red and negligible factors are greyed out.

was found in the EDAS p-side layer sequence, which contains a grin-layer with varying aluminum content right above the quantum well. In this 200 nm thick layer, the calculated local thermal resistance is twice that of the same region (no grin) in the ASLOC vertical design.

A further impact on the thermal lens profile was found by reducing the **chip width**. An initial p-up vs. p-down comparison of gain guided EDAS BALs showed that S_{th} is influenced by the distance of the active zone to the heat sink. Here, S_{th} is doubled from (0.09 ± 0.02) mm mrad/K in p-side up emitters to (0.21 ± 0.02) mm mrad/K in p-side down emitters, while BPP_0 remains unchanged. That change in S_{th} is attributed to the thermal lens shape, which shows steep edges (FEM simulation) in the case of p-down mounting and a change in gain supply for higher-order modes since p-side up devices need less current for equal ΔT_{AZ} due to the 2x larger thermal resistance.

Based on this observation, a subsequent generation of EDAS BALs was designed that featured more compact laser chips, i.e. an increased p-side thickness (d_{p-side} +42%, from 1.4 µm to 2 µm) and reduced chip width (w_{chip} -50%, from 1000 µm to 500 µm). These BALs showed a reduced slope of the center-to-edge temperature difference $\Delta T_{ce}/\Delta T_{AZ}$ (-25%) and a slight decrease in thermal lens bowing $\Delta B_2/\Delta T_{AZ}$ (-11%). As a result, the thermal slope S_{th} was reduced by 33% and the BPP_0 ground level decreased by 15% and 39%, respectively, depending on whether the active area had a double or a single quantum well.

Investigations of $w = 90\,\mu m$ BALs with deep proton implantation showed that ***lateral carrier accumulation*** (LCA) is an important factor that regulates the onset of higher order modes at increased bias. Using the implantation as a diagnostic tool, the lateral carrier profile was modified so that LCA cannot occur. For this, protons were implanted through the p-side, directly at the stripe edge ($d_{di} = 0\,\mu m$), with the n-waveguide as targeted depth, so that the quantum well exhibits a defect density at the emitter edges which is of the same order of magnitude as the injected carrier density ($\approx 1 \times 10^{18}$ cm^{-3}). By this measure, the current spreading into the regions beyond the stripe edge and the accumulation of carriers at the quantum well edges was successfully mitigated, as shown by a series of diagnostic experiments (cf. section 3.4). The resulting BALs showed an improved BPP_{lat}, which decreased from 2.6 mm mrad to 2 mm mrad at $P_{opt} = 7\,W$ due to a reduced thermal slope S_{th} (-33%). The linear radiance peaks at 3.5 W/mm mrad, which is comparable to recently reported industry emitters (cf. table 1.2). However, the implantation-induced damage compromises the conversion efficiency, which decreased by 7%-points. Nevertheless, all implanted emitters showed excellent stability of output power and beam quality in a 5000 h hard pulse (1 s on 1 s off) aging test at $P_{max} = 7\,W$.

The application of ***close index trenches***, i.e. dry etched, rectangular shaped grooves at a lateral distance $a = 5\,\mu m$ from the stripe edge was tested. The trenches used in this study exhibited $\Delta n_{eff} = 1.5 \times 10^{-3}$ and therefore induce strong lateral waveguiding. As a consequence, the BPP_0 ground level was increased by 1.5 mm mrad and the difference in S_{th} between p-side up and p-side down mounted emitters vanished, yielding a constant thermal slope $S_{th} = (0.09 \pm 0.01)$ mm mrad/K for both configurations. Furthermore, the near field profiles in trenched BALs showed steep side shoulders and were fully stabilized, as their width $w_{95\%}$ did not change with current. Hence, in devices with close index trenches, the lateral waveguide is dominated by the trench-induced effective refractive index step. In this study, a strong Δn_{eff} enabled many lateral modes even at low bias, which led to a constant BPP_{lat} increase.

A further aspect of trenches is the induced ***strain field***. After dry-etching, the trenches were covered with silicon nitride, which led to material strain originating from the trench bottom edges. It was shown that this strain produced parasitic TM-emission that exhibited very poor lateral beam quality ($BPP_{lat,TM}$ +75% compared to $BPP_{lat,TE}$ at $P_{opt} = 8\,W$). The TM near fields showed two outer lobes that were positioned directly beneath the trenches ($w_{95\%,TM}$ +5%) and the TM far fields were broadened considerably ($\Theta_{95\%,TM}$ +59%). However, since the reduction in the degree of polarization ρ' was limited to at most 6%-points (p-side down mounting), the overall impact on the total BPP_{lat} is small: $\Delta BPP_{lat} < 0.1$ mm mrad.

The assessment of ***filamentation***, i.e. self organized formation of local intensity peaks in the near field in this study was performed via a comparison of two BAL species with different modal gain Γg_0 ($11.5\,\mathrm{cm}^{-1}$ vs. $4.4\,\mathrm{cm}^{-1}$). Based on calculations in [77], this modal gain reduction yielded to a 54% decrease in filament gain γ_{max}. As a consequence, the near field contrast (i.e. strength of near field modulation - for definition, see section 3.2.3) was reduced in BALs with $\Gamma g_0 = 4.4\,\mathrm{cm}^{-1}$. However, the comparison in terms of BPP_{lat} showed no impact, so that this effect is not dominant in modern quantum well BALs that have a modal gain $\Gamma g_0 < 12\,\mathrm{cm}^{-1}$.

5.2 Results of stripe width reduction and mode filter

In chapter 4, the focus was laid on measures to reduce the number of guided modes in BALs. Two main concepts were pursued here:

First, electro-optical characterization of weakly index guided BALs with **varying stripe width** $w \in \{30, 50, 70, 90, 110\}$ µm showed the following tendencies (relative change from $w = 30\,\mathrm{µm}$ to $110\,\mathrm{µm}$ noted in brackets): With increasing stripe width w, the threshold current density j_{thr} (-37%), the series resistance R_s (-64%) and the thermal resistance R_{th} (-48%) are reduced. Simultaneously, the output power at maximum efficiency $P_{\eta max}$ (+100%) is increased. Furthermore, the efficiency decrease beyond $\eta_{c,max}$, quantified via a linear slope $Z = \Delta\eta/\Delta I|_{\eta>\eta max}$, is less steep ($\approx$ -66%) at increased stripe widths w. A full list of noted values is given in table 4.1. The BPP_{lat} measurements showed that the near field width $w_{95\%}$ grows analogous to the injection stripe width w. The far field growth rate with power $\Delta\Theta_{95\%}/\Delta P_{opt}$, however, is decreased (-66%) from $(1.9 \pm 0.3)\,^\circ\mathrm{W}^{-1}$ ($w = 30\,\mathrm{µm}$) to $(0.63 \pm 0.03)\,^\circ\mathrm{W}^{-1}$ ($w = 110\,\mathrm{µm}$). The BPP_0 ground level increases by a factor of 5.5x and the thermal slope S_{th} by a factor of 3.2x from the smallest to the widest stripe width assessed. The highest radiance was measured in BALs with narrow stripes ($w = 30\,\mathrm{µm}$) and amounts to $B_{lin} = 4\,\mathrm{W/mm\,mrad}$ at $P_{opt} = 4\,\mathrm{W}$.
In order to assess the number of guided modes, a thermal waveguide simulation was performed for different stripe widths at fixed $\Delta T_{AZ} = 15\,\mathrm{K}$. Assuming uniform mode overlap and $\Gamma_{lat,thr} = 0.9$, the resulting mode profiles were compiled to near- and far field widths and BPP_{lat} values. The number of guided modes increases at a rate of 0.16 modes per µm and the predicted BPP_{lat} deterioration rate of $\Delta BPP_{lat}/\Delta w = (0.044 \pm 0.002)\,\mathrm{mm\,mrad/µm}$ is in good agreement with the measured increase up to $w = 70\,\mathrm{µm}$. For wider stripes, the predicted far field width deviates from the measurement (2° at $w = 110\,\mathrm{µm}$), showing that here a simple uniform mode overlap cannot fully emulate the measured BPP_{lat}.

Second, a **lateral mode filter** assessed in this study uses a germanium layer that features high optical absorption and is deposited in a dry etched trench beside the injection stripe.

This *'anti-waveguide layer'* (AWL) discriminates against higher-order modes on the basis of their near field extension $w_{95\%}$. In an initial test in an ASLOC structure with shallow etched trenches ($d_{res} = 1100\,\mathrm{nm}$), the near field width was reduced by 20% and the far field width decreased by at most 2.4°, leading to a BPP_{lat} decrease of 1.5 mm mrad. In a follow-up study with EDASLOC BALs that had deeper trenches ($d_{res} = 420\,\mathrm{nm}$), a maximal linear radiance of $B_{lin} = 4.4\,\mathrm{W/mm\,mrad}$ was achieved. Here, however, the improvement in BPP_{lat} stems from a far field reduction. In both tests, the BPP_{lat} reduction was accompanied by a considerable (up to 10%-points in the EDAS structure) efficiency decrease. Furthermore, strong scattering in the data for nominally identical chips was observed, indicating an intrinsic instability of the design that strongly depends on d_{res}.

Summarized influences on S_{th} and BPP_0

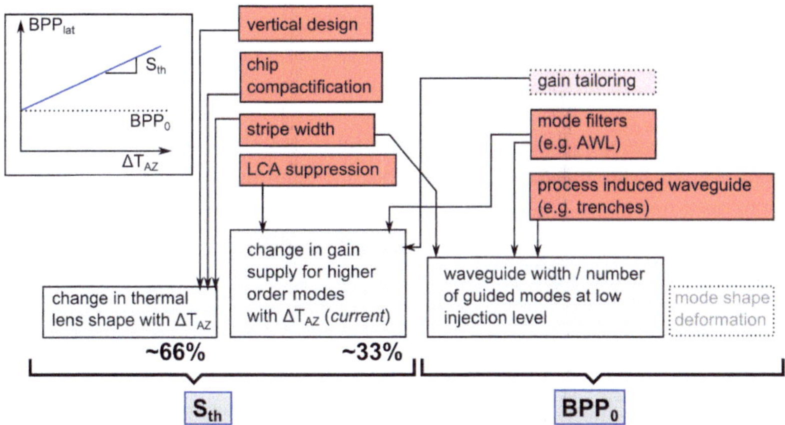

Figure 5.2: Collected influences on the BPP_0 ground level and thermal slope S_{th} from studies in this thesis. Further (suspected) influences are framed in dotted lines.

Based on the results of this thesis, the following conclusions are drawn with respect to S_{th} and BPP_0 (cf. figure 5.2). The thermal slope S_{th} is determined by two main processes. First, with increasing ΔT_{AZ}, it reflects the incremental change in the thermal lens shape, i.e. the steepness of the edges and its width, which regulates its capacity to guide lateral modes and therefore ΔBPP_{lat}. That process is estimated to be dominant, providing $\approx 2/3$ of S_{th}. The remaining $1/3$ is determined by the change in gain for higher order modes, which is a *current driven* effect that impacts S_{th} via the current dependency of ΔT_{AZ} (cf. equation 4.1).

The ground level BPP_0 is dominated by the mode capacity of the 'initial' lateral waveguide at low injection levels, i.e. the number of guided lateral modes that have equal confinement and no discrimination due to differences in modal losses. Hence, as the bias is increased, the waveguide is rapidly filled with all these modes and BPP_{lat} is elevated. A further concern, recently presented in [43], is that BPP_0 is impacted by mechanisms that deform the shape of *all* guided modes during high power operation. However, the assessment of this effect is beyond the scope of this thesis.

5.3 Outlook and design measures

From section 3.2, it is clear that, regarding the beam quality, the thermal lens is the most important issue in high power BALs. The strength of the thermal waveguide (given by ΔT_{AZ}) at a given output power P_{opt} can be reduced by enhancement of the conversion efficiency η_c and a reduction of the thermal resistance R_{th}. However, since 67% of the thermal slope is determined by the thermal lens *shape*, measures are needed to control it. The p-side of the vertical structure, the p-side metallization layers, the solder material and the heat spreader design are the promising starting points for further investigations.

Moreover, methods are sought that allow control of the BALs lateral mode spectrum without compromising its efficiency. Suppression of LCA improves S_{th} and an intrinsic discrimination mechanism for the modes in broad thermal waveguides reduces BPP_0. Here, a modification of the proton implant recipe is an obvious option if LCA suppression can be maintained without destruction of the quantum well, e.g. by a reduction in implantation dose or by adjusting the ion stop to be directly above the active area. Other methods to modify the edge regions of the active area are quantum well intermixing [93] or etch- and regrowth techniques [101]. The latter, however, must not induce a strong index guide to avoid a BPP_0 increase. In general, the use of strong index guiding trenches should be avoided, unless stable near field dimensions over a large current range are required.

Further promising approaches to access the lateral modes of a BAL are gain-tailoring, e.g. via a patterned injection stripe [68], and lateral mode filters. In this study, the AWL showed its potential in a few laser chips, but needs further optimization of the lateral design (d_{res}, dx_{AWL}, d_{Ge} etc.) for *reliable* BPP_{lat} improvement.

Appendix A

Gehrsitz-Model Refractive Index Function in MatLab

```matlab
function n = nAlGaAsGehrsitz(lambda,x,T)
AGS = (5.9613 + (0.0007178*T) - (0.000000953*T.^2)) - (16.159*x)
    + (43.511*x.^2) - (71.317*x.^3) + (57.535*x.^4) - (17.451*x
    .^5);
C0GS = 1./(50.535 - (150.7*x) - (62.209*x.^2) + (797.16*x.^3) -
    (1125*x.^4) + (503.79*x.^5));
E0GS = ((((1.5192 + (1.8*0.0159*(1 - coth(15.9./(2*0.0861708*T)))
    )) + (1.1*0.0336*(1 - coth(33.6./(2*0.0861708*T))))))))
    ./(1000000*(4.135667516e-015)*(299792458))) + (1.1308*x) +
    (0.1436*x.^2);
C1GS = 21.5647 + (113.74*x) - (122.5*x.^2) + (108.401*x.^3) -
    (47.318*x.^4);
E1sqGS = (4.7171 - (0.0003237*T) - (0.000001358*T.^2)) +
    (11.006*x) - (3.08*x.^2);
RGS = (((1 - x)*0.00155)./(0.000724 - (1./(lambda).^2))) + ((x
    *0.00261)./(0.001331 - (1./(lambda).^2)));

n = sqrt(AGS + C0GS./(E0GS.^2 - (1./(lambda)).^2) + C1GS./(
    E1sqGS - (1./(lambda)).^2) + RGS);
```

Appendix B

Transfer Matrix Calculation for Far-field Imaging

The propagation of a laser beam in geometric optics can be calculated using the transfer matrix method. Therefore the *paraxial* approximation is assumed, implying that the beams propagation angle θ with respect to the optical axis is small, such that $\tan(\theta) \approx \theta \approx \sin(\theta)$.

The laser beams path is then split into segments, each representing a particular type of propagation: free space propagation along a distance d, refraction at a thin lens with focal length f, reflexion at a mirror etc. The beam itself is represented by a column-vector $\vec{b} = \left({h \atop \theta} \right)$ that indicates the distance h and the angle θ of the beam with respect to the optical axis. Now the segments are represented by a 4x4 matrix $\hat{T}$, such that the beam after transmission through the segment is described by $\vec{b_s} = \hat{T} \cdot \vec{b}$.

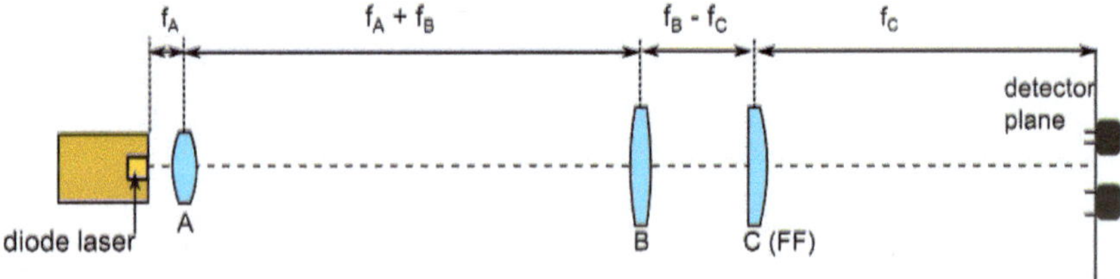

Figure B.1: Sketch of far field measurement setup in moving slit configuration. The collimating lens 'A' is placed at a distance f_A from the laser facet. The imaging lens 'B' follows at a distance $f_A + f_B$ and the far field sensor 'C' is placed at $f_B - f_C$ behind lens 'B' and at distance f_C ahead of the detector plane.

The propagation through a system of n segments with matrix representations $\hat{T}_1 \cdots \hat{T}_n$ is then simply described by multiplying the matrices in reverse order $\hat{T}_n \cdot \hat{T}_{n-1} \cdots \hat{T}_1$, yielding the system matrix $\hat{T}_{sys}$ that now can be applied to $\vec{b}$.

In figure B.1 the far field setup is presented in simplified form. For the calculation of $\hat{T}_{sys}$ it is noted, that the propagation exhibits only two types of segments, namely free space propagation, represented by $\hat{T}_d = \left(\begin{smallmatrix} 1 & d \\ 0 & 1 \end{smallmatrix}\right)$ and refraction at a thin lens, represented by $\hat{T}_f = \left(\begin{smallmatrix} 1 & 0 \\ -1/f & 1 \end{smallmatrix}\right)$.

Now the system matrix for the far field setup is obtained by the following series of matrix multiplications (starting on the right), $\hat{T}_{FF} =$

$$\begin{pmatrix} 1 & f_C \\ 0 & 1 \end{pmatrix} \cdot \begin{pmatrix} 1 & 0 \\ -1/f_C & 1 \end{pmatrix} \cdot \begin{pmatrix} 1 & f_B - f_C \\ 0 & 1 \end{pmatrix} \cdot \begin{pmatrix} 1 & 0 \\ -1/f_B & 1 \end{pmatrix} \cdot \begin{pmatrix} 1 & f_A + f_B \\ 0 & 1 \end{pmatrix} \cdot \begin{pmatrix} 1 & 0 \\ -1/f_A & 1 \end{pmatrix} \cdot \begin{pmatrix} 1 & f_A \\ 0 & 1 \end{pmatrix}$$

This multiplication results in

$$\hat{T}_{FF} = \begin{pmatrix} 0 & -\frac{f_A f_C}{f_B} \\ \frac{f_B}{f_A f_C} & -\frac{2 f_A}{f_B} \end{pmatrix} .$$

Now the propagation of a beam $\vec{b} = \left(\begin{smallmatrix} h \\ \theta \end{smallmatrix}\right)$ is simply given by:

$$\begin{pmatrix} h_{FF} \\ \theta_{FF} \end{pmatrix} = \hat{T}_{FF} \cdot \vec{b} = \begin{pmatrix} -\theta \frac{f_A f_C}{f_B} \\ h \frac{f_B}{f_A f_C} - \theta \frac{2 f_A}{f_B} \end{pmatrix} .$$

This leads to the detector position conversion given in equation 2.19, since

$$h_{FF} = -\theta \frac{f_A f_C}{f_B} \Leftrightarrow \theta = -h_{FF} \frac{f_B}{f_A f_C} .$$

List of Figures

List of Tables

Bibliography

[1] G. Overton, A. Nogee, D. Belforte, and C. Holton. Where have all the lasers gone?, Jan 2017. Annual laser market review and forecast.

[2] R. Huang. Direct diode lasers for sheet metal cutting, May 2014. Talk at Aachener Kreis Lasertechnik (AKL).

[3] F. Bachmann, P. Loosen, and R. Poprawe. *High Power Diode Lasers: Technology and Applications*. Springer Series in Optical Sciences. Springer New York, 2007.

[4] T. Y. Fan. Laser beam combining for high-power, high-radiance sources. *IEEE Journal of Selected Topics in Quantum Electronics*, 11(3):567–577, May 2005.

[5] New lab record for brilliance achieved by osram os. `https://www.osram.com/os/press/press-releases/new_lab_record_for_brilliance_achieved_by_laser.jsp`. press release 9/29/2016, Accessed: 2017-4-20.

[6] M. Kanskar, L. Bao, Z. Chen, D. Dawson, M. DeVito, W. Dong, M. Grimshaw, X. Guan, M. Hemenway, R. Martinsen, W. Urbanek, and S. Zhang. Continued improvement in reduced-mode (REM) diodes enable 272 W from 105 micrometer 0.15 NA beam. *Proc. SPIE*, 10086:1008609–1008609–6, 2017.

[7] M. Winterfeldt, P. Crump, S. Knigge, A. Maaßdorf, U. Zeimer, and G. Erbert. High beam quality in broad area lasers via suppression of lateral carrier accumulation. *IEEE Photonics Technology Letters*, 27(17):1809–1812, Sept 2015.

[8] H. C. Eckstein, M. Stumpf, U. D. Zeitner, C. Lauer, A. Bachmann, and M. Furitsch. Mode shaping for enhanced brightness in broad area lasers using monolithically integrated microoptical structures. In *2014 International Semiconductor Laser Conference*, pages 17–18, Sept 2014.

[9] H. Kissel, P. Wolf, A. Bachmann, C. Lauer, H. König, J. W. Tomm, B. Köhler, U. Strauß, and J. Biesenbach. Tailored bars at 976 nm for high-brightness fiber-coupled modules. *Proc. SPIE*, 10086:100860B–100860B–12, 2017.

[10] A. Pietrzak, R. Huelsewede, M. Zorn, O. Hirsekorn, J. Sebastian, J. Meusel, V. Bluemel, and P. Hennig. New highly efficient laser bars and laser arrays for 8xx-10xx nm pumping applications. *Proc. SPIE*, 8965:89650T–89650T–12, 2014.

[11] N.W. Ashcroft and D.N. Mermin. *Solid State Physics*. Saunders College Publishing, 1976.

[12] C. Kittel. *Introduction to solid state physics*. Wiley, 1971.

[13] L. Bergmann, H. Niedrig, and H.J. Eichler. *Optik: Wellen- und Teilchenoptik*. de Gruyter, 2004.

[14] A.E. Siegman. *Lasers*. University Science Books, 1986.

[15] S.M. Sze and K.K. Ng. *Physics of Semiconductor Devices*. Wiley, 2006.

[16] S.M. Sze. *Semiconductor Devices: Physics and Technology*. John Wiley & Sons Singapore Pte. Limited, 2012.

[17] M. Grundmann. *The Physics of Semiconductors: An Introduction Including Nanophysics and Applications*. Graduate Texts in Physics. Springer Berlin Heidelberg, 2010.

[18] G.P. Agrawal and N.K. Dutta. *Semiconductor Lasers*. Springer-Verlag US, 1993.

[19] L.A. Coldren, S.W. Corzine, and M.L. Mashanovitch. *Diode Lasers and Photonic Integrated Circuits*. Wiley Series in Microwave and Optical Engineering. Wiley, 2012.

[20] S.L. Chuang. *Physics of Photonic Devices*. Wiley Series in Pure and Applied Optics. Wiley, 2012.

[21] P. Epperlein. *Semiconductor Laser Engineering, Reliability and Diagnostics - A Practical Approach to High Power and Single Mode Devices*. Jon Wiley & Sons, 2013.

[22] R. Diehl. *High-Power Diode Lasers: Fundamentals, Technology, Applications*. Topics in Applied Physics. Springer Berlin Heidelberg, 2003.

[23] A. Einstein. Strahlungs-Emission und -Absorption nach der Quantentheorie. *Deutsche Physikalische Gesellschaft*, 18, 1916.

[24] A. L. Schawlow and C. H. Townes. Infrared and optical masers. *Phys. Rev.*, 112:1940–1949, Dec 1958.

[25] T. H. Maiman. Stimulated optical radiation in ruby. *Nature*, 187:493–494, August 1960.

[26] R. N. Hall, G. E. Fenner, J. D. Kingsley, T. J. Soltys, and R. O. Carlson. Coherent light emission from GaAs junctions. *Phys. Rev. Lett.*, 9:366–368, Nov 1962.

[27] H. Kroemer. A proposed class of hetero-junction injection lasers. *Proceedings of the IEEE*, 51(12):1782–1783, Dec 1963.

[28] I. Hayashi, M. Panish, and P. Foy. A low-threshold room-temperature injection laser. *IEEE Journal of Quantum Electronics*, 5(4):211–212, April 1969.

[29] N. Holonyak, R. Kolbas, R. Dupuis, and P. Dapkus. Quantum-well heterostructure lasers. *IEEE Journal of Quantum Electronics*, 16(2):170–186, Feb 1980.

[30] K. H. Hasler, H. Wenzel, P. Crump, S. Knigge, A. Maaßdorf, R. Platz, R. Staske, and G. Erbert. Comparative theoretical and experimental studies of two designs of high-power diode lasers. *Semiconductor Science and Technology*, 29(4):045010, 2014.

[31] E. Zucker, D. Zou, L. Zavala, H. Yu, P. Yalamanchili, L. Xu, H. Xu, D. Venables, J. Skidmore, V. Rossin, R. Raju, M. Peters, K.H. Liao, K.W. Lee, B. Kharlamov, A. Hsieh, R. Gurram, J. Guo, N. Guerin, J. Gregg, R. Duesterberg, J. Du, A. Demir, P. Cheng, J. Cheng, H. Ishiguro, R. Li, Y. Mizoguchi, and H. Sako. Advancements in laser diode chip and packaging technologies for application in kW-class fiber laser pumping. *Proc. SPIE*, 8965:896507–896507–14, 2014.

[32] M. Peters, V. Rossin, M. Everett, and E. Zucker. High-power, high-efficiency laser diodes at JDSU. *Proc. SPIE*, 6456:64560G–64560G–11, 2007.

[33] P. Crump, G. Erbert, H. Wenzel, C. Frevert, C. M. Schultz, K. H. Hasler, R. Staske, B. Sumpf, A. Maaßdorf, F. Bugge, S. Knigge, and G. Tränkle. Efficient high-power laser diodes. *IEEE Journal of Selected Topics in Quantum Electronics*, 19(4):1501211–1501211, July 2013.

[34] Martin Winterfeldt. Photonic band crystal laser. Master's thesis, Technische Universität Berlin, Germany, August 2012.

[35] W. Shockley. The theory of p-n junctions in semiconductors and p-n junction transistors. *Bell System Technical Journal*, 28(3):435–489, 1949.

[36] B. Ryvkin and E. Avrutin. Heating-induced carrier accumulation in the optical confinement layer and the output power in broadened symmetric and narrow asymmetric waveguide laser diodes. *Journal of Applied Physics*, 101(12), 2007.

[37] A. Malag, E. Dabrowska, M. Teodorczyk, G. Sobczak, A. Kozlowska, and J. Kalbarczyk. Asymmetric heterostructure with reduced distance from active region to heatsink for 810-nm range high-power laser diodes. *IEEE Journal of Quantum Electronics*, 48(4):465–471, April 2012.

[38] W.P. Dumke. Quantum theory of free carrier absorption. *Phys. Rev.*, 124:1813–1817, Dec 1961.

[39] C. Lauer, H. König, G. Grönninger, S. Hein, A. Gomez-Iglesias, M. Furitsch, J. Maric, H. Kissel, P. Wolf, J. Biesenbach, and U. Strauss. Advances in performance and beam quality of 9xx-nm laser diodes tailored for efficient fiber coupling. *Proc. SPIE*, 8241:824111–824111-10, 2012.

[40] B.J. Neubert and B. Eppich. Influences on the beam propagation ratio {M2}. *Optics Communications*, 250(4–6):241 – 251, 2005.

[41] R. J. Lang, D. Mehuys, A. Hardy, K.M. Dzurko, and D.F. Welch. Spatial evolution of filaments in broad area diode laser amplifiers. *Applied Physics Letters*, 62(11):1209–1211, 1993.

[42] P. Crump, S. Böldicke, C. M. Schultz, H. Ekhteraei, H. Wenzel, and G. Erbert. Experimental and theoretical analysis of the dominant lateral waveguiding mechanism in 975 nm high power broad area diode lasers. *Semiconductor Science and Technology*, 27(4):045001, 2012.

[43] P. Crump, M. Ekterai, C. M. Schultz, G. Erbert, and G. Tränkle. Studies of limitations to lateral brightness in high power diode lasers using spectrally-resolved mode profiles. In *2014 International Semiconductor Laser Conference*, pages 23–24, Sept 2014.

[44] M. Cross and M. J. Adams. Effects of doping and free carriers on the refractive index of direct-gap semiconductors. *Opto-electronics*, 6(3):199–216, 1974.

[45] B. R. Bennett, R. A. Soref, and J. A. Del Alamo. Carrier-induced change in refractive index of InP, GaAs and InGaAsP. *IEEE Journal of Quantum Electronics*, 26(1):113–122, Jan 1990.

[46] P. Epperlein. *Semiconductor Laser Engineering, Reliability and Diagnostics - A Practical Approach to High Power and Single Mode Devices*, chapter 2.4.2 Broad Area Lasers. Jon Wiley & Sons, 2013.

[47] S. Gehrsitz, F. K. Reinhart, C. Gourgon, N. Herres, A. Vonlanthen, and H. Sigg. The refractive index of AlGaAs below the band gap: Accurate determination and empirical modeling. *Journal of Applied Physics*, 87(11):7825–7837, 2000.

[48] M. Buda, G. Iordache, G. A. Acket, T. G. van der Roer, L. M. F. Kaufmann, B. H. van Roy, E. Smallbrugge, I. Moerman, and C. Sys. Stress-induced effects by the anodic oxide in ridge waveguide laser diodes. *IEEE Journal of Quantum Electronics*, 36(10):1174–1183, Oct 2000.

[49] J. Yang and D. T. Cassidy. Strain measurement and estimation of photoelastic effects and strain induced optical gain change in ridge waveguide lasers. *Journal of Applied Physics*, 77(7):3382–3387, 1995.

[50] R. J. Radke. *A MatLab Implementation of the Implicitly Restarted Arnoldi Method for Solving Large-Scale Eigenvalue Problems*. PhD thesis, Rice University, 1996.

[51] K. Wohlleben and W. Beck. Die Änderung von Konzentration und Beweglichkeit der Ladungsträger in GaAs bei Bestrahlung mit Protonen. *Zeitschrift Naturforschung Teil A*, 21:1057–1071, July 1966.

[52] S.J. Pearton. Ion implantation for isolation of III-V semiconductors. *Materials Science Reports*, 4(6):313 – 363, 1990.

[53] SRIM - Stopping Range of Ions in Matter. http://www.srim.org. Accessed: 2016-10-21.

[54] J. C. Dyment, L. A. DAsaro, J. C. North, B. I. Miller, and J. E. Ripper. Proton-bombardment formation of stripe-geometry heterostructure lasers for 300 K cw operation. *Proceedings of the IEEE*, 60(6):726–728, June 1972.

[55] A.G. Foyt, W.T. Lindley, C.M. Wolfe, and J.P. Donnelly. Isolation of junction devices in GaAs using proton bombardment. *Solid-State Electronics*, 12(4):209 – 214, 1969.

[56] R. W. Dixon and W. B. Joyce. (Al,Ga)As double-heterostructure lasers: Comparison of devices fabricated with deep and shallow proton bombardment. *Bell System Technical Journal*, 59(6):975–985, 1980.

[57] J. C. Dyment, J. C. North, and L. A. DAsaro. Optical and electrical properties of proton-bombarded p-type GaAs. *Journal of Applied Physics*, 44(1):207–213, 1973.

[58] A.R. Clawson. Guide to references on III–V semiconductor chemical etching. *Materials Science and Engineering: R: Reports*, 31(1–6):1 – 438, 2001.

[59] E. Colas, A. Shahar, and W. J. Tomlinson. Diffusion enhanced epitaxial growth of thickness modulated low loss rib waveguides on patterned GaAs substrates. *Applied Physics Letters*, 56(10):955–957, 1990.

[60] B. Eppich. *Landolt-Bornstein New Series, Group VIII (Advanced Materials and Technologies) Vol. 1 (Laser Physics and Applications), Sub-vol. B (Laser Systems)*, pages 145–54. Springer Verlag Berlin, 2011.

[61] Lasers and laser-related equipment - test methods for laser beam widths, divergence angles and beam propagation ratios, 2005. ISO Standard 11146-1.

[62] M. Ziegler. *Thermography of Semiconductor Lasers*. PhD thesis, Humboldt-Universität zu Berlin, 2009.

[63] M. Hempel, J.W. Tomm, F. Yue, M.A. Bettiati, and T. Elsaesser. Short-wavelength infrared defect emission as a probe of degradation processes in 980 nm single-mode diode lasers. *Laser & Photonics Reviews*, 8(5):L59–L64, 2014.

[64] F. P. Dabkowski, A. K. Chin, P. Gavrilovic, S. Alie, and D. M. Beyea. Temperature profile along the cavity axis of high power quantum well lasers during operation. *Applied Physics Letters*, 64(1):13–15, 1994.

[65] J. LeClech, M. Ziegler, J. Mukherjee, J.W. Tomm, T. Elsaesser, J.P. Landesman, B. Corbett, J.G. Mclnerney, J.P. Reithmaier, S. Deubert, A. Forchel, W. Nakwaski, and R.P. Sarzala. Microthermography of diode lasers: The impact of light propagation on image formation. *Journal of Applied Physics*, 105(1), 2009.

[66] J. Piprek. Self-consistent far-field blooming analysis for high-power fabry-perot laser diodes. *Proc. SPIE*, 8619:861910–861910–8, 2013.

[67] J. Piprek and S.Z.M Li. On the importance of non-thermal far-field blooming in broad-area high-power laser diodes. *Applied Physics Letters*, 102(22), 2013.

[68] C. Lindsey, D. Mehuys, and A. Yariv. Linear tailored gain broad area semiconductor lasers. *IEEE Journal of Quantum Electronics*, 23(6):775–787, Jun 1987.

[69] G. Agrawal. Lateral analysis of quasi-index-guided injection lasers: Transition from gain to index guiding. *Journal of Lightwave Technology*, 2(4):537–543, Aug 1984.

[70] J. R. Marciante and G. P. Agrawal. Spatio-temporal characteristics of filamentation in broad-area semiconductor lasers: experimental results. *IEEE Photonics Technology Letters*, 10(1):54–56, Jan 1998.

[71] K. I. Shigihara, K. Kawasaki, Y. Yoshida, S. Yamamura, T. Yagi, and E. Omura. High-power 980-nm ridge waveguide laser diodes including an asymmetrically expanded optical field normal to the active layer. *IEEE Journal of Quantum Electronics*, 38(8):1081–1088, Aug 2002.

[72] M. Winterfeldt, J. Rieprich, S. Knigge, A. Maaßdorf, M. Hempel, R. Kernke, J. W. Tomm, G. Erbert, and P. Crump. Assessing the influence of the vertical epitaxial layer design on the lateral beam quality of high-power broad area diode lasers. *Proc. SPIE*, 9733:97330O–97330O–9, 2016.

[73] M. Winterfeldt, P. Crump, H. Wenzel, G. Erbert, and G. Tränkle. Experimental investigation of factors limiting slow axis beam quality in 9xx nm high power broad area diode lasers. *Journal of Applied Physics*, 116(6), 2014.

[74] D. C. Hall, L. Goldberg, and D. Mehuys. Technique for lateral temperature profiling in optoelectronic devices using a photoluminescence microprobe. *Applied Physics Letters*, 61(4):384–386, 1992.

[75] Ioffe institute, thermal properties of aluminum gallium arsenide. `http://www.ioffe.ru/SVA/NSM/Semicond/AlGaAs/thermal.html`. Accessed: 2016-07-10.

[76] M.A. Afromowitz. Thermal conductivity of GaAlAs alloys. *Journal of Applied Physics*, 44(3):1292–1294, 1973.

[77] G. C. Dente. Low confinement factors for suppressed filaments in semiconductor lasers. *IEEE Journal of Quantum Electronics*, 37(12):1650–1653, Dec 2001.

[78] T. A. DeTemple and C. M. Herzinger. On the semiconductor laser logarithmic gain-current density relation. *IEEE Journal of Quantum Electronics*, 29(5):1246–1252, May 1993.

[79] J. Stohs, D. J. Bossert, D. J. Gallant, and S. R. J. Brueck. Gain, refractive index change, and linewidth enhancement factor in broad-area GaAs and InGaAs quantum-well lasers. *IEEE Journal of Quantum Electronics*, 37(11):1449–1459, Nov 2001.

[80] X. Wang, G. Erbert, H. Wenzel, P. Crump, B. Eppich, S. Knigge, P. Ressel, A. Ginolas, A. Maaßdorf, and G. Tränkle. 17-W near-diffraction-limited 970 nm output from a tapered semiconductor optical amplifier. *IEEE Photonics Technology Letters*, 25(2):115–118, Jan 2013.

[81] G. Agrawal. Lateral analysis of quasi-index-guided injection lasers: Transition from gain to index guiding. *Journal of Lightwave Technology*, 2(4):537–543, Aug 1984.

[82] K. Posilovic. *Neuartige Wellenleiterkonzepte für brillante Halbleiterlaser*. PhD thesis, Technische Universität Berlin, 2015.

[83] C. W. Higginbotham, M. Cardona, and F. H. Pollak. Intrinsic piezobirefringence of Ge, Si, and GaAs. *Phys. Rev.*, 184:821–829, Aug 1969.

[84] L. Borruel, S. Sujecki, P. Moreno, J. Wykes, P. Sewell, T. M. Benson, E. C. Larkins, and I. Esquivias. Modeling of patterned contacts in tapered lasers. *IEEE Journal of Quantum Electronics*, 40(10):1384–1388, Oct 2004.

[85] K. Petermann. Some relations for the far-field distribution of semiconductor lasers with gain-guiding. *Optical and Quantum Electronics*, 13(4):323–333, 1981.

[86] U. Zeimer, 2015. personal communication.

[87] P. M. Smowton and P. Blood. Fermi level pinning and differential efficiency in GaInP quantum well laser diodes. *Applied Physics Letters*, 70(9):1073–1075, 1997.

[88] P. M. Smowton and P. Blood. The differential efficiency of quantum-well lasers. *IEEE Journal of Selected Topics in Quantum Electronics*, 3(2):491–498, Apr 1997.

[89] W. Sun, R. Pathak, G. Campbell, H. Eppich, J.H. Jacob, A. Chin, and J. Fryer. Higher brightness laser diodes with smaller slow axis divergence. *Proc. SPIE*, 8605:86050D–86050D–9, 2013.

[90] J. Piprek. Inverse thermal lens effects on the far-field blooming of broad area laser diodes. *IEEE Photonics Technology Letters*, 25(10):958–960, May 2013.

[91] J.G. Bai, P. Leisher, S. Zhang, S. Elim, M. Grimshaw, C. Bai, L. Bintz, D. Dawson, L. Bao, J. Wang, M. DeVito, R. Martinsen, and J. Haden. Mitigation of thermal lensing effect as a brightness limitation of high-power broad area diode lasers. *Proc. SPIE*, 7953:79531F–79531F–7, 2011.

[92] Robert J. Martinsen. Method and apparatus for controlling thermal variations in an optical device, October 21 2003. United States Patent: US 6,636,539 B2.

[93] Y.R. Zhao, G. A. Smolyakov, and M. Osinski. High-performance InGaAs-GaAs-AlGaAs broad-area diode lasers with impurity-free intermixed active region. *IEEE Journal of Selected Topics in Quantum Electronics*, 9(5):1333–1339, Sept 2003.

[94] J. Buus. The effective index method and its application to semiconductor lasers. *IEEE Journal of Quantum Electronics*, 18(7):1083–1089, Jul 1982.

[95] J. Decker, P. Crump, J. Fricke, A. Maaßdorf, G. Erbert, and G. Tränkle. Narrow stripe broad area lasers with high order distributed feedback surface gratings. *IEEE Photonics Technology Letters*, 26(8):829–832, April 2014.

[96] C. Groth and G. Mueller. *FEM für Praktiker Band 3: Temperaturfelder*. Expert-Verlag, 2009.

[97] J. Rieprich, M. Winterfeldt, J.W. Tomm, R. Kernke, and P. Crump. Assessment of factors regulating the thermal lens profile and lateral brightness in high power diode lasers. *Proc. SPIE*, 10085:1008502–1008502–10, 2017.

[98] M. Yuda, T. Hirono, A. Kozen, and C. Amano. Improvement of kink-free output power by using highly resistive regions in both sides of the ridge stripe for 980-nm laser diodes. *IEEE Journal of Quantum Electronics*, 40(9):1203–1207, Sept 2004.

[99] V. P. Kalosha, K. Posilovic, and D. Bimberg. Lateral-longitudinal modes of high-power inhomogeneous waveguide lasers. *IEEE Journal of Quantum Electronics*, 48(2):123–128, Feb 2012.

[100] H. Wenzel, P. Crump, J. Fricke, P. Ressel, and G. Erbert. Suppression of higher-order lateral modes in broad-area diode lasers by resonant anti-guiding. *IEEE Journal of Quantum Electronics*, 49(12):1102–1108, Dec 2013.

[101] P. Della Casa, O. Brox, J. Decker, M. Winterfeldt, P. Crump, H. Wenzel, and M. Weyers. High power broad area buried mesa lasers, 2017. submitted to SST, not yet published.

Innovationen mit Mikrowellen und Licht
Forschungsberichte aus dem Ferdinand-Braun-Institut, Leibniz-Institut für Höchstfrequenztechnik

Herausgeber: Prof. Dr. G. Tränkle, Prof. Dr.-Ing. W. Heinrich

Band 1: **Thorsten Tischler**
Die Perfectly-Matched-Layer-Randbedingung in der
Finite-Differenzen-Methode im Frequenzbereich:
Implementierung und Einsatzbereiche
ISBN: 3-86537-113-2, 19,00 EUR, 144 Seiten

Band 2: **Friedrich Lenk**
Monolithische GaAs FET- und HBT-Oszillatoren
mit verbesserter Transistormodellierung
ISBN: 3-86537-107-8, 19,00 EUR, 140 Seiten

Band 3: **R. Doerner, M. Rudolph (eds.)**
Selected Topics on Microwave Measurements,
Noise in Devices and Circuits, and Transistor Modeling
ISBN: 3-86537-328-3, 19,00 EUR, 130 Seiten

Band 4: **Matthias Schott**
Methoden zur Phasenrauschverbesserung von
monolithischen Millimeterwellen-Oszillatoren
ISBN: 978-3-86727-774-0, 19,00 EUR, 134 Seiten

Band 5: **Katrin Paschke**
Hochleistungsdiodenlaser hoher spektraler Strahldichte
mit geneigtem Bragg-Gitter als Modenfilter (α-DFB-Laser)
ISBN: 978-3-86727-775-7, 19,00 EUR, 128 Seiten

Band 6: **Andre Maaßdorf**
Entwicklung von GaAs-basierten Heterostruktur-Bipolartransistoren
(HBTs) für Mikrowellenleistungszellen
ISBN: 978-3-86727-743-3, 23,00 EUR, 154 Seiten

Band 7: **Prodyut Kumar Talukder**
Finite-Difference-Frequency-Domain Simulation of Electrically
Large Microwave Structures using PML and Internal Ports
ISBN: 978-3-86955-067-1, 19,00 EUR, 138 Seiten

Band 8: **Ibrahim Khalil**
Intermodulation Distortion in GaN HEMT
ISBN: 978-3-86955-188-3, 23,00 EUR, 158 Seiten

Band 9: **Martin Maiwald**
Halbleiterlaser basierte Mikrosystemlichtquellen für die Raman-Spektroskopie
ISBN: 978-3-86955-184-5, 19,00 EUR, 134 Seiten

Band 10: **Jens Flucke**
Mikrowellen-Schaltverstärker in GaN- und GaAs-Technologie
Designgrundlagen und Komponenten
ISBN: 978-3-86955-304-7, 21,00 EUR, 122 Seiten

Cuvillier Verlag
Internationaler wissenschaftlicher Fachverlag

Innovationen mit Mikrowellen und Licht
Forschungsberichte aus dem Ferdinand-Braun-Institut, Leibniz-Institut für Höchstfrequenztechnik

Herausgeber: Prof. Dr. G. Tränkle, Prof. Dr.-Ing. W. Heinrich

Band 11:
Harald Klockenhoff
Optimiertes Design von Mikrowellen-Leistungstransistoren
und Verstärkern im X-Band
ISBN: 978-3-86955-391-7, 26,75 EUR, 130 Seiten

Band 12:
Reza Pazirandeh
Monolithische GaAs FET- und HBT-Oszillatoren
mit verbesserter Transistormodellierung
ISBN: 978-3-86955-107-8, 19,00 EUR, 140 Seiten

Band 13:
Tomas Krämer
High-Speed InP Heterojunction Bipolar Transistors
and Integrated Circuits in Transferred Substrate Technology
ISBN: 978-3-86955-393-1, 21,70 EUR, 140 Seiten

Band 14:
Phuong Thanh Nguyen
Investigation of spectral characteristics of solitary
diode lasers with integrated grating resonator
ISBN: 978-3-86955-651-2, 24,00 EUR, 156 Seiten

Band 15:
Sina Riecke
Flexible Generation of Picosecond Laser Pulses in the Infrared and
Green Spectral Range by Gain-Switching of Semiconductor Lasers
ISBN: 978-3-86955-652-9, 22,60 EUR, 136 Seiten

Band 16:
Christian Hennig
Hydrid-Gasphasenepitaxie von versetzungsarmen und freistehenden
GaN-Schichten
ISBN: 978-3-86955-822-6, 27,00 EUR, 162 Seiten

Band 17:
Tim Wernicke
Wachstum von nicht- und semipolaren InAlGaN-Heterostrukturen
für hocheffiziente Licht-Emitter
ISBN: 978-3-86955-881-3, 23,40 EUR, 138 Seiten

Band 18:
Andreas Wentzel
Klasse-S Mikrowellen-Leistungsverstärker mit GaN-Transistoren
ISBN: 978-3-86955-897-4, 29,65 EUR, 172 Seiten

Band 19:
Veit Hoffmann
MOVPE growth and characterization of (In,Ga)N quantum structures
for laser diodes emitting at 440 nm
ISBN: 978-3-86955-989-6, 18,00 EUR, 118 Seiten

Band 20:
Ahmad Ibrahim Bawamia
Improvement of the beam quality of high-power broad area
semiconductor diode lasers by means of an external resonator
ISBN: 978-3-95404-065-0, 21,00 EUR, 126 Seiten

Cuvillier Verlag
Internationaler wissenschaftlicher Fachverlag

Innovationen mit Mikrowellen und Licht
Forschungsberichte aus dem Ferdinand-Braun-Institut,
Leibniz-Institut für Höchstfrequenztechnik

Herausgeber: Prof. Dr. G. Tränkle, Prof. Dr.-Ing. W. Heinrich

Band 21: **Agnietzka Pietrzak**
Realization of High Power Diode Lasers with Extremely Narrow
Vertical Divergence
ISBN: 978-3-95404-066-7, 27,40 EUR, 144 Seiten

Band 22: **Eldad Bahat-Treidel**
GaN-based HEMTs for High Voltage Operation
Design, Technology and Characterization
ISBN: 978-3-95404-094-0, 41,10 EUR, 220 Seiten

Band 23: **Ponky Ivo**
AlGaN/GaN HEMTs Reliability:
Degradation Modes and Anslysis
ISBN: 978-3-95404-259-3, 23,55 EUR, 132 Seiten

Band 24: **Stefan Spießberger**
Compact Semiconductor-Based Laser Sources
with Narrow Linewidth and High Output Power
ISBN: 978-3-95404-261-6, 24,15 EUR, 140 Seiten

Band 25: **Silvio Kühn**
Mikrowellenoszillatoren für die Erzeugung von atmosphärischen
Mikroplasmen
ISBN: 978-3-95404-378-1, 21,85 EUR, 112 Seiten

Band 26: **Sven Schwertfeger**
Experimentelle Untersuchung der Modensynchronisation in Multisegment-
Laserdioden zur Erzeugung kurzer optischer Pulse bei einer Wellenlänge
von 920 nm
ISBN: 978-3-95404-471-9, 29,45 EUR, 150 Seiten

Band 27: **Christoph Matthias Schultz**
Analysis and mitigation of the factors limiting the effiency of high power
distributed feedback diode lasers
ISBN: 978-3-95404-521-1, 68,40 EUR, 388 Seiten

Band 28: **Luca Redaelli**
Design and fabrication of GaN-based laser diodes for single-mode
and narrow-linewidth applications
ISBN: 978-3-95404-586-0, 29,70 EUR, 176 Seiten

Band 29: **Martin Spreemann**
Resonatorkonzepte für Hochleistungs-Diodenlaser
mit ausgedehnten lateralen Dimensionen
ISBN: 978-3-95404-628-7, 25,15 EUR, 128 Seiten

Cuvillier Verlag
Internationaler wissenschaftlicher Fachverlag

Innovationen mit Mikrowellen und Licht
Forschungsberichte aus dem Ferdinand-Braun-Institut, Leibniz-Institut für Höchstfrequenztechnik

Herausgeber: Prof. Dr. G. Tränkle, Prof. Dr.-Ing. W. Heinrich

Band 30: **Christian Fiebig**
Diodenlaser mit Trapezstruktur und hoher Brillanz für die Realisierung
einer Frequenzkonversion auf einer mikro-optischen Bank
ISBN: 978-3-95404-690-4, 26,30 EUR, 140 Seiten

Band 31: **Viola Küller**
Versetzungsreduzierte AlN- und AlGaN-Schichten
als Basis für UV LEDs
ISBN: 978-3-95404-741-3, 34,40 EUR, 164 Seiten

Band 32: **Daniel Jedrzejczyk**
Efficient frequency doubling of near-infrared diode lasers
using quasi phase-matched waveguides
ISBN: 978-3-95404-958-5, 27,90 EUR, 134 Seiten

Band 33: **Sylvia Hagedorn**
Hybrid-Gasphasenepitaxie zur Herstellung von Aluminiumgalliumnitrid
ISBN: 978-3-95404-985-1, 38,00 EUR, 176 Seiten

Band 34: **Alexander Kravets**
Advanced Silicon MMICs for mm-Wave Automotive Radar Front-Ends
ISBN: 978-3-95404-986-8, 31,90 EUR, 156 Seiten

Band 35: **David Feise**
Longitudinale Modenfilter für Kantanemitter im roten Spektralbereich
ISBN: 978-3-7369-9116-3, 39,20 EUR, 168 Seiten

Band 36: **Ksenia Nosaeva**
Indium phosphide HBT in thermally optimized periphery for
applications up to 300GHZ
ISBN: 978-3-7369-287-0, 42,00 EUR, 154 Seiten

Band 37: **Muhammad Maruf Hossain**
Signal Generation for Millimeter Wave and THZ Applications
in InP-DHBT and InP-on-BiCMOS Technologies
ISBN: 978-3-7369-9335-8, 35,60 EUR, 136 Seiten

Band 38: **Sirinpa Monayakul**
Development of Sub-mm Wave Flip-Chip Interconnect
ISBN: 978-3-7369-9410-2, 44,00 EUR, 146 Seiten

Band 39: **Moritz Brendel**
Charakterisierung und Optimierung von (Al, Ga) N-basierten
UV-Photodetektoren
ISBN: 978-3-7369-9465-2, 49,90 EUR, 196 Seiten

Band 40: **Erdenetsetseg Luvsandamdin**
Development of micro-integrated diode lasers for precision quantum optics
experiments in space
ISBN: 978-3-7369-9479-9, 39,00 EUR, 126 Seiten

Cuvillier Verlag
Internationaler wissenschaftlicher Fachverlag

Innovationen mit Mikrowellen und Licht
Forschungsberichte aus dem Ferdinand-Braun-Institut,
Leibniz-Institut für Höchstfrequenztechnik

Herausgeber: Prof. Dr. G. Tränkle, Prof. Dr.-Ing. W. Heinrich

Band 41: **Thi Nghiem Vu**
Development and analysis of diode laser ns-MOPA systems
for high peak power application
ISBN: 978-3-7369-9480-5, 38,80 EUR, 138 Seiten

Band 42: **Christian Bansleben**
Differentieller Mikrowellen-Leistungsoszillator für die Realisierung
ultrakompakter Plasmaquellen in Matrixanordnung
ISBN: 978-3-7369-9530-7, 34,90 EUR, 136 Seiten

Cuvillier Verlag
Internationaler wissenschaftlicher Fachverlag